The Science of Education

Part 2B: The Awareness of Mathematization

Caleb Gattegno

Educational Solutions Worldwide Inc.

First published in the United States of America in 1988. Reprinted in 2010.

Copyright © 1988– 2010 Educational Solutions Worldwide Inc.
Author: Caleb Gattegno
All rights reserved
ISBN 978-0-87825-208-4

Educational Solutions Worldwide Inc.
2nd Floor 99 University Place, New York, NY 10003-4555
www.EducationalSolutions.com

Acknowledgements

The manuscript of this volume was handed to my friend of many years, Dick Tahta, so that he makes out of it a book that teachers of mathematics at all levels may find readable and interesting. This is what was done and I want to express my gratitude and that of readers of this work for the result. The preparation of the text for the printers involved Ms. Lisa Wood who also did Part 1 and Dr. D.E. Hinman whose experience in making books attractive through titles and subtitles has been proved once more in this volume. To both go my thanks and appreciation for their devotion and care.

Table of Contents

Preface

The three chapters which make up this second part of the applications of The Science of Education are concerned with a new look at mathematics, a venerable subject cultivated from antiquity by many civilizations. Because mathematics is recognizable but not easily defined, we replaced it by a process or processes which can be made more tangible and that we named "mathematization." Hence the general title of this volume, The Awareness of Mathematization.

Until now, it was left to chance to produce the number of mathematicians the world needed. Now we can work deliberately and diligently at increasing their numbers: we now know how to give everybody the opportunities which go into providing the criteria that will permit anyone of us to say, "I like occupying my time with these types of mental exercises, and I will go on pursuing these lines of activities." Each individual will be able to decide whether becoming a professional mathematician suits him or her, or whether, after tasting this intellectual nourishment, they prefer other involvements to become more permanent. Such a choice was never mentioned in the past because the gift for doing math was considered rare and not easily brought about in those who did not display it spontaneously.

This volume attempts to show how at least the first steps can be taken in this enterprise.

* * *

No one marvels at the fact that most young children everywhere learn to speak, not knowing that the demands of this apprenticeship are much greater than those of learning school mathematics.

If we do marvel, and study first language acquisition, we discover that human minds are equipped to perform mental acts which resemble what mathematicians do in their own mental laboratories. It follows that a new track is available to the scientist of education concerned with making mathematicians: to begin with language, to become aware of its unfolding, and to get as much help as possible from it in order to specialize some mental structures already at work so as to give them some of the mathematical attributes.

Chapter 10 is there to show how.

Once all these mental mathematical structures are made plain and are systematically approached, other awarenesses can be used to increase the crop and make it integrate with the elementary school math curriculum. Chapter 11 is given to this task.

Though we gather a lot in the pages of these two chapters, we can notice that the writer left out other attributes of the mind which are routinely and systematically used by mathematicians in their research and are visible in their findings. These attributes were left out as a device of exposition in order to ensure mastery of what was being presented and not because their importance was being ignored. Chapter 12 shows that they can be taken up separately and that they do more than give more entries into mathematics. The attributes belong to the education of the whole brain seen as the servant and ally of the mind. Under that lighting much more mathematics is encountered, acknowledged, known and enjoyed, but still with the same qualities of being spontaneously invented by each student who plays the games in which he or she is given that little which is unlikely to be uncovered, but leads to a lot. That lot which covers years of study in the traditional curriculum, requires now only a few months.

Saving time, giving enjoyment, opening great vistas all follow from this daily encounter with this awareness of mathematization. As a product of The Science of Education it deserves to be widely known and studied for the sake of the next generation.

Caleb Gattegno
June 1988

10 Beginning Mathematics

Introduction

This chapter of The Science of Education sets out a curriculum for elementary mathematics and describes ways of forcing awareness in students, which can meet all the demands considered legitimate for 5- to 12-year-olds all over the world, and which, at the same time, can be the source of a working knowledge of mathematics that would satisfy the professional point of view of any mathematician.

With this approach, people will find they can learn as naturally as in their other spontaneous learning, acquire skills which can be recalled at once, and become mathematicians in the field they study. This remarkable revolution has been made possible by some essential discoveries of The Science of Education. Some of the most important of these findings are briefly outlined here.

- Children who start school mathematics have already reached a level of mental evolution which is illustrated by, among other things, their considerable mastery of spoken language.

- Many activities of children require an organization of mental operations which can only be described as algebraic in nature.

- Underlying the notations of mathematics there are verbal components; so the mastery of the spoken language means that it is possible to base mathematics on language.

- Every arithmetical problem has a background of algebra. This represents awareness of the process, while the numerical aspect represents the concrete particular that is given, so there is far less algebra to assimilate than there are numerical relations to take into account. Hence algebra should be stressed whenever possible.

- Traditions of teaching — curricula, methods, textbooks, workbooks, and so on — have found their way into schools and into teacher education on the basis of authoritatively expressed opinion, without proper investigation into what mathematics is, and whether there are ways of working that can offer mathematics to all; in particular, whether "mathematization" is not for working mathematicians only, but open to all who attempt it in a disciplined manner.

- Bright ideas are not enough — though they may help sometimes. What is needed instead is a subordination of teaching to learning and a critical examination of What children do with their minds in their spontaneous activities — like learning to speak — as well as a study of what it is in traditional teaching that plays such a great part in children's failure in mathematics.

An illustration of the importance of stressing algebra appears in the development of the use of colored rods. Georges Cuisenaire showed in the early fifties that students who had been taught traditionally, and who were rated "weak," took huge strides when they shifted to using the material. They became "very good" at traditional arithmetic when they were allowed to manipulate the rods.

It was then demonstrated — in a number of countries in the late fifties — that children's progress came from the fact that the rods act as an algebraic model — for the algebra of arithmetic — making it possible to start with algebra instead of counting. It made sense to the students. They paid attention to perceptible attributes and had very little to memorize. Therefore they could re-invent, easily and on every occasion, what was needed to solve a problem, and did not worry they might forget facts only held in their memory.

When it was found that the rods act as an algebraic model, it became possible not only to improve teaching by starting with algebra, but also to look for other "foundations of mathematics" than were available in the literature under the prestigious names of Frege, Peano, Cantor, Hilbert, and so on. These other foundations were found where no one was looking: in the universal ability of babies to acquire the language of their environment.

To see mathematics learning as based on the powers used by young children to acquire language can make an even greater impact on mathematics teaching than was achieved with the rods. For these told us nothing of the learning powers of children. Treated as teaching aids, they had been linked to a priori theories like those of Montessori or Piaget. No such theory was required when it became obvious that children had the adequate mental powers to solve such a complex human problem as learning the mother tongue. Once these powers were known, it was merely a matter of working out a new approach, that brought to young children's awareness what, precisely, was demanded in order to transform their spontaneous mental structures into mathematical mental structures.

Over the last twenty years, this new approach has been constantly chiseled and refined into a <u>complete</u> curriculum, with methods of work that are easily adaptable, for any class of children, by teachers everywhere. The rest of this chapter will outline that proposal.

Language of Numeration

In books on grammar, numerals are adjectives. It is therefore legitimate to consider numeration as part of language before looking at its mathematical aspects.

Every language treats numeration differently in its choice of which words, or combination of words, have to be used to describe how many objects there are in a given set. In this section, we are not concerned with the process of attributing a numeral to a given set; this will be studied later. What we start with here is a problem of language: <u>naming the numerals</u> in any one language, so that they can become part of the vocabulary of that language, to be learned and retained like all other words. Once this has been mastered, it will be available for other uses; in particular, as a foundation for mathematics.

Different languages have various ways of naming numerals. That there is no universal way of doing this tells us that each culture had a choice in the matter. Thus it is a linguistic challenge and not a mathematical one. Since there is not just one way of naming numerals, readers may have to examine for themselves how to adapt the following treatment of English numeration to suit another language. This has already been done for some languages in the numeration charts that form part of The Silent Way method of teaching foreign languages.

The first task is to find a way to bring students to the mastery of ordinary numeration in the language used to teach them at school. This needs to be done in systematic steps to secure full retention, with complete understanding and in the shortest time. This means using the smallest number of those units of energy, required to retain anything arbitrary, which we have called <u>ogdens</u>.

What may be less obvious is that this challenge must be taken strictly for what it is: learning to read and write numerals in the language being used and not yet for other purposes such as counting. We will then be free to be guided by the required learning, rather than by other considerations which will find their place later. Since we shall be choosing a particular unfolding of the subject matter, this may seem to be an absolute. But there are variables involved — for example, the order in which certain names are introduced — which permit alternatives within that absolute. Readers should look out for such variations and decide which exposition they prefer.

* * *

For the English language, we choose to start by ensuring that the ogdens are paid to master the one-digit numerals. There are nine of them in the common system.

1   2   3   4   5   6   7   8   9

Whether they are introduced in order or not, the thing that matters is that any one of these signs triggers one sound — the traditional one — and conversely. Nine ogdens are involved in that (perhaps eighteen, if the converse is taken to require a distinct retention). Up to half an hour may be required for mastery. After that, one more ogden will be required to generate a new set of signs.

100 200 300 400 500 600 700 800 900

These are written by placing the sign "00" on the right of each unit sign, and named by saying the word "hundred" after each unit sound. We can now generate eighty-one more numerals by stringing in order any one of the hundreds followed by any one of the units — for example, "six hundred three," or "one hundred nine" — and by agreeing to telescope the two signs into a single one; for example, 600, 3 into 603, or 100, 9 into 109. Conversely, any such string with a 0 in the middle will trigger the required sound. When made explicit, these conventions are easily accepted and practiced.

So far we have demanded ten ogdens (or twenty, if spoken and written forms are counted separately — readers are asked to make this possible distinction for themselves in further cases that arise). For the cost of these ogdens, <u>we force awareness</u> that not everything that is learned has to be memorized — only that which costs ogdens has to be. Moreover there is room for initiative on the part of learners who therefore discover they can be inventive in this field of study and produce much more than what they were asked to memorize. Hence they will <u>know</u> the new numerals, rather than <u>remember</u> them through drill and repetition. Furthermore, behind the "condensed writing" of three-digit numerals is something that will soon become part of the "place-value" aspect of numeration.

When making a simultaneous display of the sequences introduced so far, a space is left between the rows of units and hundreds — without explanation, because none is needed. This empty row will then be filled in as follows. The sign 40 is written between 4 and 400 and named "forty;" 60 is now written between 6 and 600 and, if it is not forthcoming, named "sixty;" similarly for 70, 80, 90. As well as pairing numerals from two rows as previously described, it is now possible to use the three rows to produce three-digit numerals, provided we use the rows in order starting with the hundreds row.

1	2	3	4	5	6	7	8	9
			40		60	70	80	90
100	200	300	400	500	600	700	800	900

So — in English — just one more ogden, for the "ty," yields many more numerals in both written and spoken forms; for example, 69, 123, "forty one," "one hundred eighty two," and so on. To contrast three-digit numerals with a 0 in the middle or on the right, it is enough to test a few pairs — e.g., 104 and 140 — by naming them and asking for the written sign, or conversely. It only requires a few examples to be given for all the many numerals now available to be invented confidently by analogy. Such creativity, using only the few ogdens students were asked to pay in order to retain what they could not have invented themselves, constitutes a remarkable yield. Moreover, this also enables "place-value" to be met, practiced and become second-nature.

To complete the numeration up to 999 we need to introduce and integrate the "irregular" names that complete the middle row. Thus, on writing 50 between 5 and 500, this is not named "fivety" but "fifty;" on writing 30 and 20, these are named "thirty" and "twenty." Each of these requires one new ogden but nothing else.

The final entry, 10, in the middle row requires special treatment in English. The name "ten" is only used on its own or when paired with numerals from the hundreds row; for example, 10, or 110, 210, . . . , 910. When "ten" is paired with the units, there are three cases requiring the forcing of new, distinct awarenesses. In the pairings 14, 16, 17, 18, 19, the order must be reversed, with the units named as usual, but with the "ten" becoming "teen" — for example, "fourteen". The reversal is a special awareness, but these pairings still only require the one ogden for the "teen." The pairings 13, 15, are named similarly, but using the forms for 3, 5 that were used in the middle row; so, "thirteen," "fifteen." Finally, for 11, 12, we need two new ogdens to produce the names "eleven," "twelve."

The integration of "ten" in English numeration has been shown to require four new ogdens — for "ten," "teen," "eleven," and "twelve" — with the acceptance that all the other teens start with the unit numerals, with 3 and 5 being altered by "borrowing" the ogdens paid for use in 30 and 50. In all, we have generated 999 numerals for the cost of 18 ogdens.

* * *

Once it is possible to read and write three-digit numerals, it is very easy to extend that power, since the language of numeration requires but one further ogden to allow us to read and write numerals of up to six digits by calling a space, or a comma, to the left of the hundreds digit, "thousand;" for example, 123 456 — or 123, 456 — read as "one hundred twenty three thousand four hundred fifty six." Similarly "million" and the American "billion" will be the names — requiring two further ogdens — for spaces or commas on the left of the sixth and ninth digits. This extends the power of reading and writing numerals to strings of up to twelve digits for a total cost of 21 ogdens.

Note that the units row contained nine digits but that to write all the numerals an extra digit, namely 0, was used to which we have not yet had to give a name. It is "zero," but is sounded as a silence when it occurs in a string of digits' for example, 1,000 is read "one thousand," and 1,024 is read "one thousand twenty four." Moreover, a string of three zeros preceding a comma causes that comma not to be sounded; for example, 1,000,000 is read "one million." Practice with various examples containing zeros will help to make these conventions clear, they do not require further ogdens. It may be necessary to force the awareness that by including 0 among the one-digit numerals we have all we need to write, and hence read, any string of digits.

* * *

At this stage, we have the option of introducing the notion of a base of numeration. The array of three rows can be used to test whether such a notion is accessible to the students. If a transversal line is drawn between two adjacent columns of numerals in the array, then all that was done before can be repeated, but using only the part of the array to the left of the transversal and leaving out altogether the numerals on the right.

1	2	3	4	5	6	7	8	9
10	20	30	40	50	60	70	80	90
100	200	300	400	500	600	700	800	900

It is this restriction in the use of the elements of the array to produce three-digit numerals — and hence, with commas, strings of greater length — which generates the language of numerals in various bases. The reading and writing of numerals in any base requires no more than has already been done.

To name the base in terms of the common system, we select the first unit to the right of the transversal; for example, with the transversal lying between 4 and 5, the restricted array will yield numerals in base 5. We can extend the numeration system by adding further signs — e.g., letters — to the row of units. If the sign after 9 is chosen to be X, we could also conveniently say — with the Romans — that our common system is in base X.

* * *

Some of the advantages of starting our teaching of mathematics to young children with language, with the reading and writing of numerals, may be listed as follows.

- Triggers are used to obtain sounds or written signs and we know exactly how many do the whole job; these triggers join the many that children have used for years when learning their mother tongue.

- The stress on language frees us to lengthen the strings of digits, at almost no extra cost, and to eliminate the feeling that these produce more difficult notions, beyond what children can grasp at this age; strings do not become "harder," just "longer."

- The array used to generate numerals implicitly contains from the start the variable we call the base of numeration; this can be derived in a very simple way.

- The approach can be carried out in any ordinary classroom with a chalkboard on which the array can be developed as described; a pointer can be a useful way of involving the students in using the array themselves.

Numerals in Order

It may not have been noticed by readers familiar with the sequence of integers that none of the work of the last section required any notions other than those presented. For instance, the numerals we introduced did not need an <u>order structure</u> to be present. Therefore, if we want order we must introduce it deliberately.

All of us, in our schooling, have met the sounds of the "integers" or "whole numbers" embedded in a sequence in which order plays a part from the start; this is "counting" as a verbal exercise. The order can be made explicit in a development of the array used in the previous section. In using this array, whatever order in which elements were named was not a feature of the presentation, even if the standard order was used. By choosing this standard order and making it explicit, we force awareness that it can be created. We may even give it a name, calling it an order "from left to right;" another order can be indicated as easily and named an order "from right to left." Later we can call these "ascending" and "descending" orders, but these names imply other notions which will become clear in later sections.

We can now produce a two-dimensional array in which the first 99 numerals will appear in ten rows and ten columns. We start with a first row of units, with a blank in the first position; the first column is then continued with the "tens," 10 to 90. These elements can be paired as in the previous section, though here this can be done systematically to complete the new array.

·	1	2	3	4	5	6	7	8	9
10	11	12	13	14	15	16	17	18	19
20	21	22	23	24	25	26	27	28	29
30	31	32	33	34	35	36	37	38	39
40	41	42	43	44	45	46	47	48	49
50	51	52	53	54	55	56	57	58	59
60	61	62	63	64	65	66	67	68	69
70	71	72	73	74	75	76	77	78	79
80	81	82	83	84	85	86	87	88	89
90	91	92	93	94	95	96	97	98	99

An order is imposed on this array, namely the order of reading from top to bottom, each row being read from left to right, as in the usual reading of English text. No new ogdens are required for this — only practice. For example, starting at any numeral in the array we can either go "forward" to 99 or "backward" to 1, saying or touching <u>everyone</u> of the numerals in the sequence. Of course, we are helped by the awareness of a certain regularity in the naming and can give ourselves clues to know which place belongs to which of the 99 numerals.

When children can recite such sequences, we say that they can <u>count</u>. There are usually two meanings of counting; it should be born in mind that one being introduced here is part of language training. This "ordinal counting" further prepares children to enter mathematics. It is considered separately from the reading and writing of numerals because mathematicians have taken order to be a separate, basic structure of their science.

Nothing new is required to complete the ordering of numerals up to 999; it only takes more time, with the next hundred rows taking up more space. But there is no need to spend long on this, since it involves just an extension of a well-established awareness. In this case practice may be experienced as tedious, and therefore not to be imposed. Once the three-digit numerals have been ordered, it is a simple matter to extend the ordering to the larger numerals introduced in the previous section.

Counting

In this section, we address the second meaning of counting — the one that is demanded by the question "how many?" We engage in this aspect of counting when we have in front of us sets of discrete objects, which can be perceptibly separated, and when we see that the activity of counting does not go on indefinitely but stops at a certain stage.

When a "finite set" is given and we are asked how many elements there are in it, we use the following notions.

- We own the first, ordinal meaning of counting — the unfolding in order of the set of numerals.

- We can touch, point at or otherwise attend to, the elements of the set of objects in turn.

- We can synchronize two actions: attending to elements of the set and, at the same time, uttering one of the numerals in order, starting with "one." The shifting from one spoken numeral of the sequence to the next needs to coincide with our shifting attention from one of the objects to another.

- All the objects are attended to — each of them once and only once.

The last numeral uttered when the last object has been attended to is called the <u>cardinal</u> of that set. This is the answer to the question: "How many objects in this set?" A cardinal is also a numeral; all numerals can become cardinals, when they are associated with sets in the way described above, but they can be considered on their own as in the previous two sections.

We can find certain properties of cardinals by performing the action of counting the elements of various sets and observing what happens. Thus, it turns out that it is immaterial which element of the set we choose to start counting from; all would do equally well. It is also immaterial which element the counting ends on. This means that any given set will have one single cardinal associated with it. We can say that cardinals associated with sets are "invariant" with respect to the different ways of finding them; they are independent of the order of executing the counting.

Of course, many sets may have the same cardinal, in which case we say that they have "as many" elements or that they are equivalent. For example, in general most of us have as many fingers as each other and as many toes; we use this fact all the time when counting on our fingers.

Because we can vary the base of numeration when counting orally, we can become aware that there are several ways of naming the cardinal of a set. For example, the same set of fingers on our two hands may have the following cardinals: 10, 11, 12, 13, 14, 20, 22, 101, 1010, according as the last digit on the left of the transversal is 9, 8, 7, 6, 5, 4, 3, 2, 1, respectively. Though this is an important awareness, it does not need to be pressed on students who only need to have at the back of their minds the notion that there are bases of numeration and that they affect the naming of cardinals.

Complementarity

Traditionally the operations of addition and subtraction are sharply distinguished. When a written algorithm is introduced, each is treated in two stages: without, and then with, carrying in the case of addition; without, and then with, borrowing in the case of subtraction. Although numerals are always read and written from left to right, as is usual with English text, addition and subtraction are often carried out from right to left — hence, the complications of carrying and borrowing.

As soon as we ask whether there is a common way of looking at these two operations, a common source for their definition, we find that complications disappear, that some traditional methods have only been used because it was not known how to handle addition and subtraction simultaneously. Clearly it is reasonable to unify this field of elementary mathematics if this can be done. It also makes children's work much easier, more under their control and more quickly covered.

The main perception required for such unification is that any set can be subdivided into two subsets which taken together make up the original set and which are such that neither contains an element of the other. Each subset is then called the <u>complement</u> of the other within the original set. Complementarity is the new awareness we want to force on beginners. It will be a common source for the subsequent treatment of addition and subtraction.

* * *

We can consider complementarity very easily on the set of fingers of our two hands, using <u>folding</u> as the action which subdivides the original set into two separate subsets. In the usual base of numeration, the set of our fingers has ten elements. But the fingers are different from each other. This tells us that for the work we shall be doing here there is not yet a need for the notion of a "unit;" this will be introduced later to yield new awarenesses that can provide a foundation for mathematics.

Meanwhile, if we fold one or more fingers in turn, we can generate subsets, whose cardinals can be linked in statements which are immediately obvious. For example, if we fold one of our fingers and count the elements in the two subsets so formed, we find that 9 goes with 1, or 1 goes with 9. If we use some formal notation to translate our perceptions into statements, we may move toward a way of describing exactly what we have perceived; such a process is a hallmark of mathematics. Thus perception of the above complementarity may be expressed as (9, 1) or (1, 9). Similarly, we find that 8 goes with 2 or 2 goes with 8, (8, 2) or (2, 8); and so on.

$$(1, 9) \quad (2, 8) \quad (3, 7) \quad (4, 6) \quad (5, 5)$$
$$(9, 1) \quad (8, 2) \quad (7, 3) \quad (6, 4)$$

These pairs describe complements in the complete set of fingers in which none are folded. This "none" will be given the sign 0, which has already been called "zero." Hence we now have two further pairs: (10, 0) when no fingers are folded, and (0, 10) when all fingers are folded.

With the awareness that the whole set is present in each pair, we can say that these pairs display <u>all</u> the possibilities of complementarity on the set of fingers. We can say that these pairs are <u>different</u> ways of showing that the set of fingers can be replaced by two complementary subsets, each pair indicating how many fingers are up and how many are down. On the other hand, we shall also say that all pairs are <u>equivalent</u>, using the sign~, read as "equivalent to" or "is another way of saying."

$$(0, 10) \sim (1, 9) \sim (2, 8) \sim (3, 7) \sim (4, 6) \sim (5, 5)$$
$$\sim (10, 0) \sim (9, 1) \sim (8, 2) \sim (7, 3) \sim (6, 4)$$

With these eleven equivalent pairs, found on one's hands through the operations of folding or unfolding fingers, we encounter the basic notion of <u>ordered pairs</u>, from which elementary arithmetic can now be developed.

* * *

One first development is in the awareness that just as there is no obstacle to giving the name of "finger" to each of our fingers, we could name them with any one of the words we met when we moved from one row of the array of numerals to the next by introducing "hundred" or "thousand." For example, using "hundred" as the name for each finger, the full set would be called "ten hundred" and the subsets "one hundred," "two hundred" and so on. This yields eleven equivalent pairs, such as

(100, 900), (700, 300) and so on. Similarly, using "ty" as the name for each finger, the full set would be "tenty" and the subsets "onety," "twoty" and so on, yielding eleven further equivalences, such as (40, 60), (50, 50) and so on.

Of course, there are other names in English for "onety," "twoty" and so on. We are here introducing arithmetical notions by using a uniform language. This involves the introduction of a few special, but regular words; but we can also remind students that the standard names are used in other contexts.

We can create further cardinals by using more than one pair of hands at a time. For example, the fingers of one pair of hands are named "hundred" and those of another pair are named "ty." The pairs are then linked by substituting one finger of the first pair — say, a thumb — by a set called "tenty," represented by the second pair of hands. The four hands are held up in a row, with one thumb of the first pair permanently folded in order to concretize its substitution by the second pair; the full set can now be read as "nine hundred tenty."

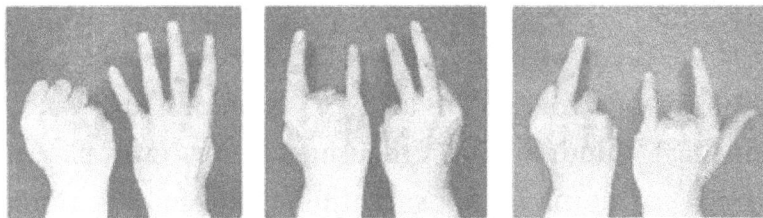

Alternative ways of showing *four*.

Two ways of showing *forty-seven*.

Complementary subsets may now be produced by folding some of the fingers. The names of these subsets can be read off easily; for example, "six hundred sixty" for the fingers that are up, "three hundred forty" for those that are down. These complements, in the "nine hundred tenty" we started with, could be expressed as (660, 340).

A third pair of hands could be held up and linked with two others in a similar way. Thus, permanently folding a thumb of the middle pair and substituting for it the set "ten" represented by the third pair, provides a complete set "nine hundred ninety ten." Here, too, by folding some fingers we nay generate two complementary subsets which can be labeled with their respective cardinals; for example (628, 372).

The actual activity consists in perceiving how many fingers are up — or down — on each pair of hands in turn. Starting with the furthest to the left and moving right, look first at those which are up, name the respective cardinal and write down its corresponding string of digits. Then do the same for those fingers which are down. We can only fold up to nine fingers on each pair of hands, other than the pair on the furthest right, because these pairs were denied one thumb in order to be linked with the pair on its right. So, the complements within each of these pairs will be *complements in nine.*

$$(0, 9)\sim(1, 8)\sim(2, 7)\sim(3, 6)\sim(4, 5)$$
$$\sim(9, 0)\sim(8, 1)\sim(7, 2)\sim(6, 3)\sim(5, 4)$$

It is not necessary to memorize these complements in nine, or those in ten. Any one of the pairs generates all the others by changing one finger at a time from being up to being down, or conversely. It is important to make sure beginners have this awareness; it helps them be independent of memory and puts the stress on perception — which does not require the payment of ogdens.

* * *

Another development of the above awareness will provide a way of immediately writing down the complement of a cardinal of any number of digits in a cardinal which is written with that number of zeros after a 1. For example, to find the complement of 723,451 in 1,000,000, we scan each digit from the left, writing down its complement in nine, except for the last digit on the right for which we write its complement in ten — giving in this case 276,549; as another example, note the following collection of complements in 100, arranged in a two dimensional array.

(0, 100)~(1, 99)~(2, 98)~(3, 97)~ . . .
(10, 90)~(11, 89)~(12, 88)~(13, 97)~ . . .
(20, 80)~(21, 79)~(22, 78)~(23, 77)~ . . .

. . .

Teachers following the curriculum outlined so far will need to stress the forcing of the awarenesses involved; the advantage in doing this will become clearer in the next section which is devoted to addition and subtraction. It will also be necessary to determine with students how much practice each of them will need in order to make complementarity second nature, before continuing further.

Addition and Subtraction

By starting from language, we can give students a thorough acquaintance with the set of numerals, expressed in the common, or any other, base. Language can be further extended through counting and complementarity to the state where we can take the important step of introducing numbers, associated with an algebra — a first entry into mathematics as such.

Numerals are words because they appear in our awareness as adjectives. They gain the property we called "cardinal" through the simultaneous awareness that their association with sets of any objects whatsoever will now always be present. So, cardinals are here understood to be numerals associated with discrete sets. To generate <u>numbers,</u> in the sense that mathematicians use, we need to add new awarenesses. These come from another look at complementarity, to see now something that was always there — but which we had previously deliberately postponed in order to be able to force awareness more thoroughly.

It is clearly a shift of awareness which tells us that in a given set of fingers on two hands with all, some, or none, of the fingers folded, then none, some, or all, respectively, will be unfolded. We can see the two subsets as well as the whole set, even if the perceptions of these are qualitatively different. Now if we perceive the two subsets within the whole set, we can say that together they make up the whole set, and we can call this perception <u>addition</u>. To make this more concrete, we introduce a new notation, replacing the bracketed pairs with which we previously named the complementary subsets. The sign + is used to characterize our new awareness and from now on will be the sign for addition.

$$10 \sim 1+9 \sim 2+8 \sim 3+7 \sim 4+6 \sim 5+5$$
$$\sim 9+1 \sim 8+2 \sim 7+3 \sim 6+4$$

We can now translate all the work done in the section on complementarity in terms of addition. But there is another awareness just as readily available, when we see the folded fingers as having been "taken away" from the whole set. In this case, we stress the relationship of the folded subset to the whole set. This awareness clearly differs from that of addition; we call it <u>subtraction</u>. From the whole set we "subtracted" the folded subset, which is no longer perceived as part of the whole set.

* * *

Although addition and subtraction are such different awarenesses of the <u>same</u> situation, they are linked by having a common source, that of complementarity. Hence, we can say that it is only a matter of awareness which lets us stress addition or subtraction in a given situation.

For instance, when we look at the two subsets, we may notice that in naming the pair we have a choice, whether to start with the folded, or with the unfolded, set. We can also <u>feel</u> that there is no reason to start with one rather than the other. Once such <u>indifference</u> becomes a focal awareness, it can be labeled; we say — adapting a technical term used by mathematicians — that addition is <u>commutative</u>. This awareness is expressed in writing as A+B ~ B+A, where A, B are the names of the two subsets and our indifference is apparent in the sign ~, which here too can be read as "is another way of saying."

Of course, this is no longer the case when we stress the awareness that leads to subtraction. For then we contrast the set of folded fingers with the whole set; and here there is <u>no</u> choice. Thus, all beginners can understand that we cannot say that subtraction is commutative: because it is not. But there is another aspect to which we can draw attention. If we write A for the folded subset, B for the unfolded one, and X for the whole set, then B also represents "what is left of X when A is taken away." We may express this awareness as B ~ X-A. In view of our indifference with respect to the two subsets that make up X, the last statement may also suggest another, A ~ X-B.

Our awarenesses so far can be summarized by the following expressions.

$$\sim A+B \text{ or } X \sim B+A \text{ and } A+B \sim B+A$$
$$\sim X-B \text{ or } B \sim X-A$$

These can be practiced as much as may be necessary, using the cardinals of the sets involved; for example, from 10~7+3 we can also derive 10~3+7, 7~10-3, 3~10-7, and so on.

This way of working eliminates altogether the need for drill in order to memorize the so-called facts of addition or subtraction. When students understand how the notation follows their perception, and how the cardinals used are linked to each other, they will know that they can <u>generate</u> the statements they make, instead of <u>recalling</u> them. Independence and autonomy of this sort breeds responsibility.

* * *

Our study of complementarity, and of the two "operations" of addition and subtraction, used our hands because they are "handy" and are never forgotten or left behind. But if we shift to any set of objects — such as dots on paper, grains of rice, or any other separable things which can be counted — we can still easily force awareness of the whole set and the many, many ways it can be subdivided into two complementary sets.

It is this awareness which is translatable into general relationships where numerals are replaced by letters. These letters are not "generalized" numerals — although one day they may appear as such. Their appearance in written expressions serves to shift the focus from the variable numerals to the operations involved and thus to force awareness of the <u>algebra</u> behind these relationships.

Work with students can now stress either the general <u>mental dynamics</u> which produces new awarenesses, or some particular examples obtained by specializing the relationships involved. Students will then be able to generate for themselves complete sets of particular cases — for instance, the specific tables containing "facts" of addition or subtraction — rather than receive them as knowledge given by the teacher. It may be that the students need not actually write down the complete sets. For mathematics is of the mind: all that is required is an understanding how particular facts could be produced when necessary. Our students then become mathematicians, rather than performers of feats of memory.

Numbers

As a further illustration of what can now be easily derived from the new foundations we have laid, consider all the equivalent pairs of complements in various cardinals that we could generate, as we did for complements in 10; for example, the following complements in 9, 8, and so on down to 2.

$$(9, 0) \sim (8, 1) \sim (7, 2) \sim (6, 3) \sim (5, 4)$$
$$(8, 0) \sim (7, 1) \sim (6, 2) \sim (5, 3) \sim (4, 4)$$
$$(7, 0) \sim (6, 1) \sim (5, 2) \sim (4, 3)$$
$$(6, 0) \sim (5, 1) \sim (4, 2) \sim (3, 3)$$
$$(5, 0) \sim (4, 1) \sim (3, 2)$$
$$(4, 0) \sim (3, 1) \sim (2, 2)$$
$$(3, 0) \sim (2, 1)$$
$$(2, 0) \sim (1, 1)$$

And all these read backwards.

Once such sets of pairs are generated, they can be integrated into existing relationships to produce a new awareness. Using the notation for addition introduced above, we can write down all the ways in which a particular cardinal, say 4, can be expressed as an addition of two others. For example, 3+1; but now note that the 3 in this last expression can be replaced by the equivalent addition 2+1. This yields a further expression 2+1+1 equivalent to 4. These equivalent additions are called <u>partitions</u>, in this case, of 4.

As an illustration of our new awareness, consider the *complete set of partitions* of each cardinal up to 5. This will be sufficiently typical to allow students, who so wish, to tackle the more cumbersome sets for larger cardinals. The cardinal 1 has only itself as a partition. The cardinal 2 has two partitions: itself and 1+1. The cardinal 3 has four partitions: 3, 2+1, 1+2, and replacing the 2 by its partition 1+1, gives the further partition 1+1+1. We can then find eight partitions of 4, and sixteen partitions of 5, in the same way.

1

2   1+1

3   2+1   1+1+1   1+2

4	3+1	2+1+1	1+1+1+1
	2+2	1+2+1	
	1+3	1+1+2	
5	4+1	3+1+1	2+1+1+1 1+1+1+1
	3+2	1+3+1	1+2+1+1
	2+3	1+1+3	1+1+2+1
	1+4	1+2+2	1+1+1+2
		2+1+2	
		2+2+1	

Each of these complete sets of partitions forms a new entity which we shall call <u>number</u>. We will use this name to mean that we are aware of a class of equivalent additions — an *equivalence class with respect to addition* — providing a wealth of relationships which can be invoked according to requirements.

When we write the sign "5" we are concerned with a <u>numeral</u> in isolation. When we associate this numeral with sets, then 5 becomes a <u>cardinal</u>, the cardinal of the sets. When we are sure that 5 can trigger any one of its partitions at any time, then it has become a <u>number</u>, sometimes more specifically called a <u>whole number</u>, or an <u>integer</u>.

In this usage, integers are abbreviated ways of referring to complicated entities. <u>Integers are classes</u> — complete sets of partitions. These classes are generated successively by replacing partitions in previous classes with equivalent additions, in <u>all</u> possible ways — as was illustrated above. Note that the replacements double the number of partitions in each successive complete set.

Thus, for integers, much that is implied remains latent; we are in a new state of mind, one in which <u>potential relationships</u> are now implied. There is no need to make these relationships explicit, only to have access to all of them and to be able then to go to the one that is needed for any particular challenge. This state of mind is the one which characterizes mathematicians. So we have achieved a lot for students through the little that has been treated in this chapter. That lot, and that little, are both awarenesses, as are the relationships between them.

* * *

Partitions are based on addition; we now consider what happens when we allow subtraction to be in our minds at the same time. From every addition of the form A+B ~ X, we can produce two subtractions, A ~ X-B and B ~ X-A. Hence from our set of partitions, which are equivalent additions, we can generate <u>equivalent subtractions</u>. For example, from 3~2+1 we can derive l~3-2 and 2~3-1, and so on <u>indefinitely</u>.

1~2-1~3-2~4-3. . .~98-97. . .~1754-1753 . . .

2~3-1~4-3. . .~75-73. . .~1261-1259. . . . . . .

3~4-1~5-1. . . . .

. . .

Tables of partitions, however large, remain limited — we can say <u>finite</u>. Sequences of pairs of subtractions never end; it is customary to describe this by saying they are <u>infinite</u>, although it seems more proper to call them <u>indefinite</u> at this stage of awareness. Any integer now implies an indefinite equivalence class of subtractions as well as a finite equivalence class of additions.

What matters here is that we have increased enormously the contents of the minds of our students, without fear of confusion, because we have been so careful to distinguish the separate items generated by awareness of awarenesses. The foundations of mathematics lie in awarenesses which eventually will become "axioms" and "logical terms." We do not try to generate an axiomatic system here because we already have its equivalent in terms of awarenesses, which are as complex for the beginner as they are for sophisticated working mathematicians. It is a separate, particular task to replace the living dynamics filling the minds of learners by a formal, rigid, <u>a priori</u> system, which would be valued for cultural reasons and not for any light it sheds on the working of the mind when doing mathematics.

* * *

We shall devote the rest of this section to seeing how to make students certain that when asked to add or subtract a pair of integers, they know precisely what to do, and are able to do it swiftly and with no chance of error when they stay with the problem.

To get the answer to a problem — such as "add *a* to *b*" or "subtract *a* from *b*" — we have to take the following steps:

- Inspect the problem to know exactly what it asks for;

- Refrain from giving an "answer," unless this is obvious and immediately forthcoming — as, for example, when finding complements;

- <u>Transform</u> the problem by the means available — equivalences — into a form from which the "answer" can be read at sight.

Since <u>numbers</u> are involved, we are simultaneously aware of equivalence classes of additions and subtractions. <u>Transformation through equivalence</u> alters the given form until it yields an answer by immediate inspection. Thus, when the given form is an addition, we shift from one of the numbers to be added — called an <u>addend</u> — a component which would make up, with the other addend, a number with as many zeros as possible. For example, for "876+649" we choose a shift from the second term that, with the first term, will make up 900; this shift — of 24 — yields the form "900+625" and by another obvious shift this becomes "1000+525," from which we can read the answer 1525.

There are many other possible transformations we could have chosen. We do not have to choose the ones that merely suit <u>our</u> tastes or prejudices. Inspection of the given pair will determine whether both the numbers have to be changed or none — as would be the case when they are already obvious complements of some number; experience of such complementary pairs helps in choosing useful changes when these have to be made. Addition can then be done from the left, so that the numbers involved can remain wholes, without the "carrying" that is required for addition from the right.

In the case of subtraction, there are two possibilities open to us. In the first, we alter the pair of numbers by adding or subtracting the same amount to both, so that the larger number has only zeros after its left-hand digit. The answer is then read as the complement of the smaller number in this latter. For example,

$$2031\text{-}784 \sim 2247\text{-}1000 \quad (\text{both } +216)$$
$$\sim 1247$$

or 2031-784 ~ 2000-753   (both -31)

　　　　　　　~ 1247

The second possible transformation for subtraction is to alter the pair by seeing the larger number as made up of two addends, one with just zeros after its left-hand digit and the other made up of the remaining digits. The answer is then read by adding this remainder to the complement of the smaller number in the one with zeros. For example,

　　2031-784 ~ 2000+31-784

　　　　　　~ 1216+31

　　　　　　~ 1247

All of these transformations can be written out, of course, in vertical form; this only needs some further attention to be given to a way of indicating what operation is involved. For example,

$$
\begin{array}{cccc}
2031 & 2000\ +31 & & \\
-784 \rightarrow & -784 & \rightarrow & 1216 \\
\hline
& \overline{} & & +31 \\
& & & \overline{1247}
\end{array}
$$

Multiples and Reciprocals

In this section, we consider the awarenesses that those who have assimilated the contents of the previous sections will need when considering the multiplication of integers. The following section will complete this task, as well as bringing division into the same framework.

When introducing the integers, it was then not necessary to understand a certain property, which we now want to bring to the reader's attention. Since numerals are words, there is not much meaning in saying that any one of them is made up of "equal units." In the order structure induced on the numerals, it makes sense to say that 8 comes <u>after</u> 7 — when counting forward from 1, or even that 7 comes after 8 — when counting backward. But it does not make sense to say that 8 is "larger" than 7, or that 7 is "smaller" than 8.

But when we move from numerals to cardinals, so that sets are present, the word "after" can, indeed, be replaced by "larger" — and "before" by "smaller" — because the elements the numerals refer to are actual objects. Though these do not have to be <u>equal</u> — for fingers, grains of rice, apples and so on, are all different — spatial perception assures us that there are "more" objects in a set of 8 than in a set of 7 — and "fewer" when we reverse the statement. This means, moreover, that we can only subtract a smaller number from a larger one. When we write $a - b$, we imply that *a* is "greater than" *b,* which is written *a > b,* or that *b* is "smaller than" *a,* written *b < a.*

Passing to integers is only possible if we introduce the notion of a <u>unit</u> and require that the units in all partitions are equal — or <u>replicas</u> of one of them. So we can have an indefinite number of 1's, all equal in the sense that they are indistinguishable from each other, except for the order in which they appear in any written form. Similarly, we can have an indefinite number of 2's, of 3's and so on. These enable us to write expressions such as 1+1, 2+2+2, 3+3+3 and so on.

This awareness of the existence of equal units gives numbers some properties that are not called for when we use numerals or cardinals as such. We shall, therefore, be able to do more with numbers than with numerals or cardinals — for example, involve them in algebraic operations.

* * *

Let us call the set of integers from 1 onwards (I). Although the following work can be done in any base of numeration, we shall limit our discussion to the common system. The set (I) can be generated as a sequence of successive additions of one unit to the previous term. Such additions would soon become cumbersome to express; for example the fourth term would have to be written as ((1+1)+1)+1. We could indicate the generation of the sequence more easily by using an arrow $\rightarrow$ with the sign "+1" to mean <u>add</u> 1 and writing the results of successive applications of this operation.

$$(I) \quad 1\xrightarrow{+1}2\xrightarrow{+1}3\xrightarrow{+1}4\xrightarrow{+1}5\xrightarrow{+1}\ldots$$

Here and elsewhere, the three dots indicate that the sequence carries on and on. The indefinite set (I) can be written simply, and more succinctly, as follows.

$$(I) \quad 1 \ 2 \ 3 \ 4 \ 5 \ 6 \ 7 \ 8 \ 9 \ 10 \ 11 \ldots$$

From (I) we can derive another sequence, called the set of <u>odd</u> integers and labeled (O), by starting with 1 and skipping alternate terms.

$$(O) \quad 1 \ 3 \ 5 \ 7 \ 9 \ 11 \ 13 \ 15 \ 17 \ 19 \ldots$$

The complement of this set in (I) is called the set of <u>even</u> integers and labeled (E).

$$(E) \quad 2 \ 4 \ 6 \ 8 \ 10 \ 12 \ 14 \ 16 \ 18 \ldots$$

Both (O) and (E) have the property that any successive pair of integers differ by two units — units which remain equal for successive pairs; thus both could be generated by successive additions of two units, starting from 1 and 2, respectively.

If we compare (E) and (I) we find that for <u>every</u> element *a* in (I) there is a <u>corresponding</u> element in (E) which is equivalent to *a* +*a*. We shall also write this as 2, calling it the <u>double</u> of *a,* or <u>twice</u> *a,* or <u>two</u> *a,* so that the set of even integers is seen to be <u>also</u> made up of the doubles of the elements of (I); so that we shall also write (E) as (2I).

The shift of awareness from (E) to (2I) gives an entry to new relationships which can be kept in mind, jointly and separately, according to the demands of any particular challenge. In this way, we enrich our minds without burdening our memory.

By placing (I) and (2I) beneath each other we can display the doubling relationship, which is now to be singled out for special consideration. The relationship can be expressed by a <u>downward</u> arrow with the sign ×2, read <u>multiplied by 2;</u> the elements of (2I) are called <u>multiples of 2</u>. These expressions will also be used later with other numbers.

(I)	1	2	3	4	5	6	7	...
	↓x2	↓x2	↓x2	↓x2	↓x2	↓x2	↓x2	
(2I)	2	4	6	8	10	12	14	...

We can now look at any number and its double, to discover that their numerals display a property which helps us find the double of any integer directly without having to unfold the whole sequence from the start. Thus, the double of 324 is 648, where the double may be found by doubling the digits separately. There are, of course, many other examples of this sort; but there are some numbers that cannot be doubled in this way and a new awareness is needed for these.

This new awareness takes us back to what we did when we telescoped <u>numerals</u> as 600, 70 and 3 into the form 673. In the case of <u>numbers</u> we create a similar effect by <u>adding</u> the component numbers. So that we can double the number 673 by considering the more easily doubled components 600, 70 and 3.

$$673 \times 2 \sim (600+70+3) \times 2$$
$$\sim 600 \times 2 + 70 \times 2 + 3 \times 2$$
$$\sim 1200+140+6$$
$$\sim 1340+6 \text{ or } 1200+146$$
$$\sim 1346$$

We can find the double of any number, in this way. But note that when we write a string of more than two addends — as in 600+70+3 — we must distinguish sharply a <u>convenience</u> in writing and the <u>necessity</u> to perform the operation of addition step by step. To express the actual operation in writing that makes sense requires brackets, and brackets within brackets;

this can become tedious and inconvenient as in the following expressions

$$(\, (\, (\, (3+4)+5)+6)+7) \; \sim \; (3+(\, (4+5)+(6+7) \,) \,)$$

These expressions give a complex, though correct, account of how the additions are to be performed. But where we are not concerned with the particular order in which the step by step additions are to be carried out we may, by an "abuse of language," dispense with brackets. The sense of an abuse may be alleviated by an <u>agreement</u> not to keep strictly to <u>writing</u> every addition just with two addends, although this is the way we shall always perform them. We express this agreement by saying that addition is an <u>associative</u> operation.

* * *

Once we can understand how to double any integer, one possible development, involving a new awareness, is to consider the <u>inverse</u> of doubling, which we call <u>halving</u>; we shall also refer to this by saying "find half of" or "multiply by $\frac{1}{2}$," and by writing "$\times \frac{1}{2}$".

$$(I) \quad 1 \qquad 2 \qquad 3 \qquad 4 \qquad\qquad \ldots$$

$$\uparrow \times \tfrac{1}{2} \quad \uparrow \times \tfrac{1}{2} \quad \uparrow \times \tfrac{1}{2} \quad \uparrow \times \tfrac{1}{2}$$

$$(II) \quad 2 \qquad 4 \qquad 6 \qquad 8 \qquad\qquad \ldots$$

It may be helpful to ask beginners to look again at the sequences (I) and (2I). Instead of moving from the elements of (I) to the corresponding ones in (2I), we start with an element of (2I) and ask for the corresponding element of (I). We can look at x2 and $\times \dfrac{1}{2}$ — which are called <u>reciprocals</u> of each other — as <u>operators</u>, capable of <u>doing</u> something and which are seen to have a very precise meaning. Here the doing is to move from one sequence to the other, "downward" or "upwards." The two operators are intimately related, so that neither is harder than the other; it can become second nature to use them simultaneously.

Exercises can be given at three levels.

- Select any element of (I) or (2I) and determine the corresponding one by either doubling or halving. For example, double 756, halve 1024.

- Select two elements, one in (I) and one in (2I), and determine their relationship; this may involve an addition or a subtraction, followed by a doubling or a halving according to the choices made, synthesizing all the previous learnings. For example, relate 206 and 152 — for which one

48

solution might be "add 98 to the 206 to make 304 and then halve this."

- Select only <u>one</u> element, either in (I) or in (2I), and express its relationship to an element of the other sequence, given only by its position in it. This exercise provides the foundation of the solution of linear equations on integers. For example, find the number that is "twice 17, add 6."

Such exercises can be made very attractive to young children, particularly when presented as microcomputer games, where the verbal part of our earlier discussion can be advantageously ignored — for example, by animating the screen as in the Visual and Tangible Maths software (cf disc 11, modules 2, 3, 4).

* * *

Another possible development is to <u>iterate</u> operations like doubling or halving by repeating them over and over again. This does not entail a new awareness but does provide us with new material on which to work. We double the sequence (I) to obtain (2I). If we iterate the doubling we pass from the sequence (2I) to a sequence we can call (4I), and then to (8I). We do not need to go beyond this in terms of what is needed for elementary school mathematics, but — of course — there is no end to the further iterations we could make.

$$(I) \quad 1 \quad 2 \quad 3 \quad 4 \quad 5 \quad 6 \quad 7 \quad \ldots$$
$$(2I) \quad 2 \quad 4 \quad 6 \quad 8 \quad 10 \quad 12 \quad 14 \quad \ldots$$
$$(4I) \quad 4 \quad 8 \quad 12 \quad 16 \quad 20 \quad 24 \quad 28 \quad \ldots$$
$$(8I) \quad 8 \quad 16 \quad 24 \quad 32 \quad 40 \quad 48 \quad 56 \quad \ldots$$

Once these sequences are available, further practice, with the kind of exercises that have been suggested for work with the first two, will provide students with much more experience of <u>multiples</u> and their <u>reciprocals</u> — $\times 2$, $\times \frac{1}{2}$; $\times 4$, $\times \frac{1}{4}$; $\times 8$, $\times \frac{1}{8}$ — and with <u>equations</u>, to yield a feel for numbers rarely offered to people at this stage. Although the important progress lies in the algebra behind all this, it is at the same time also possible to cover such bits of knowledge as the "multiplication tables" — so far, for 2, 4, and 8 — which are considered important in elementary schools.

	1	2	3	4	5	6	7	8	9	10	11	12
x 2	2	4	6	8	10	12	14	16	18	20	22	24
x 4	4	8	12	16	20	24	28	32	36	40	44	48
x 8	8	16	24	32	40	48	56	64	72	80	88	96

When a number a in the first row of the above array is multiplied by any number b in the first column, we shall say that $a \times b$ is a <u>multiplication</u> that yields a number called a <u>product</u> — this will be located at the crossing of the relevant row and column. The 36 products in the above table can be looked at in three ways, and these can be exercised, for example, by students generating their own challenges in pairs.

- $7 \times 8 \sim ?$, reading an answer 56 at a crossing;

- $? \times 4 \sim 36$, reading an answer 9 in the first row;

- $6 \times ? \sim 48$, reading an answer 8 in the first column.

This yield is restricted compared to the considerable other yields found so far, but it is the one that tests of achievement are often concerned with; here, it is easily obtained in passing.

Note that in our usage so far, $a \times b$ is "*b* times *a*." But this term of (*b*I) is also the result of successive additions of *b*, namely "*a*-times *b*," which is represented by $b \times a$, hence $a \times b \sim b \times a$, a symmetry that may already have been noticed — for example in $2 \times 8 \sim 16 \sim 8 \times 2$. We express this symmetry by saying that multiplication is commutative. In our usage, this means that multiples of *a* are also multiplications <u>by</u> *a*.

* * *

Doubling was chosen first to illustrate iteration because of its fecundity. The next stage in the study of multiples and reciprocals can be chosen because of its facility.

Multiples of 10 and, indeed, of all the numbers written with the digit 1 followed by a number of zeros, are easily obtained. Any number multiplied by ten yields that number of tens, which is written by following the digits of the number with one zero; for example, $27 \times 10 \sim 270$. Iterating the operation $\times 10$ a number of times then simply means writing in that number of zeros. Examples can be worked out first through writing and then by reading these numerals — $6 \times 1000 \sim 6,000$ or "six thousand," $35 \times 100 \sim 3,500$ or "three thousand five hundred," and so on.

The iteration of operations like $\times 2$, $\times \frac{1}{2}$, $\times 10$ or $\times \frac{1}{10}$ present the same algebra, called <u>exponentiation</u>, which is a new awareness that can be developed in a separate branch of arithmetic with its own further unfolding.

But what we want to do here is to enable students to develop the way they look at the problems they meet so that they learn to solve each of them by the simplest, <u>ad hoc</u> approach. For example, since we know that $10 \times \frac{1}{2} \sim 5$, we can replace every multiplication by 5 by the two operations $\times 10$ and $\times \frac{1}{2}$, that have been already mastered. Such an algebraic transformation of a given problem immediately reveals a remarkable extension of awareness: any number with only even digits multiplied by 5 yields a new number whose digits are obtained by halving the given ones and placing a zero to the right of the string. For example, 2466×5 ~ 123340, and there is nothing intimidating in treating longer strings of even digits — these will just take more time. The algebra has yielded an indefinite number of multiples of 5, all as easily obtained as the above example.

When the digits of a number are a mixture of even and odd, a scrutiny of the sequence may help in finding the form of its multiplication by 5. Since an even number can be seen as a double, it may be represented in the form $2a$. Therefore, an odd number may be represented by a form $2a+1$; hence an odd number multiplied by 5 will end with a 5. Asking students to see an odd number as "an even number +1" will make them see that any odd number multiplied by 5 is obtained by halving the even number and placing a 5 on the right; for example 73×5 ~ (72+1)×5 ~ 72×5 + 5 ~ 360+5 ~ 365. This now means that all multiples of 5 can be obtained.

(5I)   5 10 15 20 25 30 35 40 45 50 . . .

* * *

Next, we consider multiples of 3, which can be found by adding twice an integer to that integer. Since we already have the sequences (I) and (2I) we can by a simple addition obtain the sequence (3I). Doubling this yields (6I), and doubling again (12I).

$$(3I) \quad 3 \quad 6 \quad 9 \quad 12 \quad 15 \quad 18 \quad 21 \quad 24 \quad 27 \quad 30 \quad ...$$
$$(6I) \quad 6 \quad 12 \quad 16 \quad 24 \quad 30 \quad 36 \quad 42 \quad 48 \quad 54 \quad 60 \quad ...$$
$$(12I) \quad 12 \quad 24 \quad 36 \quad 48 \quad 60 \quad 72 \quad 84 \quad 96 \quad 108 \quad 120 \quad ...$$

Two further sequences of multiples of 9, and of 11, may be obtained by taking elements of (I) and subtracting these from, or adding them to, the corresponding elements in (10I).

$$(9I) \quad 9 \quad 18 \quad 27 \quad 36 \quad 45 \quad 54 \quad 63 \quad 72 \quad 81 \quad 90 \quad ...$$
$$(11I) \quad 11 \quad 22 \quad 33 \quad 44 \quad 55 \quad 66 \quad 77 \quad 88 \quad 99 \quad 110 \quad ...$$

Exercises can be carried out with these rows like the ones proposed for previous ones — these could include work on the reciprocals $\times \frac{1}{3}$, $\times \frac{1}{6}$, $\times \frac{1}{9}$, $\times \frac{1}{11}$, and $\times \frac{1}{12}$.

To complete the rows of multiples up to multiples of 12, we need finally to generate the multiples of 7. We could work out the indefinite set (7I) by using (I) and (6I). But, instead, we could draw students' attention to the seventh entry in each of the established rows which will contain some <u>multiplications</u> of 7 — for example, 7×3~21. Since multiplication is commutative these will also be the required <u>multiples</u> of 7 — for example, 3×7.

$$\times 7 \quad 7 \quad 14 \quad 21 \quad 28 \quad 35 \quad 42 \quad . \quad 56 \quad 63 \quad 70 \quad 77 \quad 84$$

We can insert the missing entry by calculating 7×7 as 6×7 +7 or 8×7 -7, either of which yields 49. This completes the traditional "7-times table."

* * *

Indeed, we now have all the elements of the so-called Pythagorean table of products up to 12×12. Producing this is a very minor, almost trivial, part of the rich mathematical yield gained by using the sequence (I) and its successive multiples, obtained through doubling and other operations. The mathematics we have done replaces the need for memorization — so prevalent in elementary schools — by a mental activity that is always at the level of the students and which they can fully comprehend.

The introduction of "reciprocals," or "equations," and of exercises which sharpen the mathematical eye of students, enormously increases the range and depth of the traditional curriculum. That we have been able to do this is an indication that emphasizing awareness is so much more effective than any approach that ignores it altogether.

Multiplication and Division Algorithms

It has been implicit in some of the calculations given in the previous section that numbers can be expressed as additions of certain components which are written as elements from the array of numerals in the first section. This allows us to see any number as an ordered sequence of products, each made up of a one-digit number multiplied by "a power of 10;" for example 351 ~ 3×100 + 5×10 + 1. The phrase "a power of 10" is another way of describing a number that is written with a 1 followed by a string of zeros. Thus, instead of 1000, which has three zeros, we shall also say "third power of 10" or "10 to the power three," and write "10^3". A few exercises will show that all this is quite straightforward; for example 1000 ~ $10^?$, 10^2 ~ ?, 10^1 ~ ?, and 10^0 ~ 1, the last being an extrapolation from the others, one that can be considered for the present to make sense <u>purely as a notation</u>.

A sequence of additions of multiples of powers is called a <u>polynomial</u>. To obtain the <u>polynomial form</u> of any number we take the corresponding numeral and read it as a sequence of elements from the rows of the array of numerals; for example, 9745 is read as 9000, 700, 40, 5. We then rewrite the string using + signs, and then using powers of ten.

9745 is read "9000, 700, 40, 5"

$$9745 \sim 9000 + 700 + 40 + 5$$
$$\sim 9{\times}10^3 + 7{\times}10^2 + 4{\times}10^1 + 5$$

We have already used this polynomial form surreptitiously in the previous section — for example, when doubling a number by doubling digits separately. It is now required explicitly in order to be able to multiply any integer by another. This involves a new awareness which mathematicians have called the <u>distributive law</u>.

In effect, there is also a second awareness involved — the inverse of the first — called <u>factoring</u> or <u>factorization</u>. It is important not to postpone this inverse operation, because by stressing language and algebra, we can treat all inverse operations simultaneously — as we have already done in the case of addition and subtraction, or multiples and reciprocals. But as there is no need for the inverse in the present context, we concentrate here on the distributive law.

There are various approaches we can take. We can use colored rods, as has been shown in various specialized publications, such as Mathematics with Numbers in Color, books 1-6 — further details are given in the appendix. Where the appropriate facilities are available, we can use computer graphics, as in Visible and Tangible Mathematics; disc 12, program 12.2, and this is preferable because the forcing of awareness is then easier. In either case, we represent a product by the area of a rectangle;

but note that we use area only to provide an intuitive support for some algebraic awarenesses — we do not take the notion of area to be a primitive one in mathematics.

With the rods, we start by making a "train" of rods of different colors to convey at once that one of the dimensions of the rectangle that will represent a product is a sum of lengths. Placing a certain number of replicas of the train side by side widens the rectangle, making it more noticeable. This may then be seen as made up of smaller rectangles, all having one common dimension which may be represented by the length of a rod.

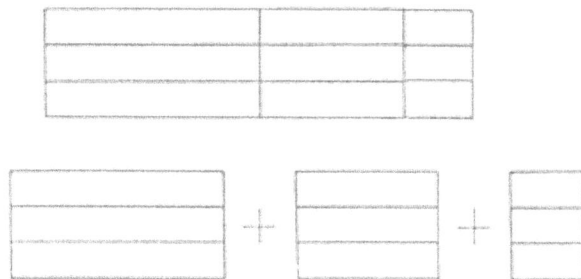

Denoting the common dimension by p and the lengths of the rods in train by a, b, c, and so on, our perception can be written as follows.

$$p \quad \times \quad (a+b+c) \sim p \times a \; + \quad p \times b \; + \quad p \times c$$

On the left, the sign + means an addition of lengths, while on the right, it has the meaning of adding areas. This need not be confusing, since both lengths and areas are numbers in the work we are doing here. It is clear that p has been "distributed" to a, b, and c; this tells us why the distributive law has been so named. On the other hand, a reverse algebraic reading may be seen as the selection of a "common factor" p.

If we use more than one rod to represent the second dimension — the width — of our large rectangle of rods, then we replace p by, say, $p+q$, and the products on the right of the above equivalence can then be replaced "by distribution."

$$(p+q) \times (a+b+c) \sim p \times a \ + \ q \times a \ + \ p \times b \ + \ q \times b \ + \ p \times c \ + \ q \times c$$

A similar approach may be taken using computer graphics. In either case, "distribution" is seen to replace an original product that may be difficult to calculate by a string of products that may be able to be calculated more easily.

* * *

Of course, "long" multiplications may nowadays be carried out less tediously with calculators. But if we do not have a calculator — or wish to do without one for some reason — we can still use a two-dimensional array — so-called "vertical notation" — which helps to process the distributed products involved.

Traditionally, operations in arithmetic have been performed from the right. But the reading and writing of number polynomials — and the action of the distributive law — are from the left. So we first develop a routine for multiplication from the left; and then, to comply with tradition, we rearrange the array so that students can <u>also</u> work from the right. The various stages involved — illustrated by a worked example — are listed below.

- Being aware of distribution, carried to in its <u>horizontal</u> form, from left to right, as shown above.

746×35 $\sim (700+40+6) \times (30+5)$

$\sim 700 \times 30 + 700 \times 5 + 40 \times 30 +$ etc

$\sim 21000 + 3500 + 1200 + 200 + 180 +$ etc

- Translating into <u>vertical</u> form (see below), with the calculated partial products aligned and ordered, starting with the largest, namely the first on the left in the horizontal form.

- Minimizing the number of partial products by adding all those having the same number of zeros on the right.

- Adding the partial products to obtain the required single product.

```
      746            746            746
    x  35          x  35          x  35
    21000          21000          21000
     3500   →              →
     1200           4700           4700
      200
      180            380            380
       30             30             30
                                   26110
```

- Re-writing the whole procedure by reversing the vertical order of the partial products; this is equivalent to now multiplying from the right.

- Abbreviating the written calculation by ignoring all zeros on the right that indicate the power of ten in each partial product, but making sure that the remaining, so-called "significant," digits — which may include zeros — are aligned correctly.

746	746	746
x 35	x 35	x 35
30	30	30
180		
200	380	38
1200		
3500	4700	47
21000	2100	21
		26110

This routine — called the <u>multiplication algorithm</u> in the traditional curriculum — can be practiced with students until it is clear what the steps are and what each of them means, particularly the last abbreviation.

* * *

We may consider division as the inverse process of multiplication. For any given integer n, called a <u>divisor</u>, we ask what number it may be multiplied by to yield a given, larger, integer N, called a <u>dividend</u>. The answer — called the <u>quotient</u> — is also said to be the number of times the divisor n "goes into" the <u>dividend</u>. In general, the quotient is not an integer, but we shall start by choosing n and N so that it is.

To develop a division algorithm we force awareness of a new reading of sequences of multiples. Consider (I), the standard set of integers, and (nI), the set of multiples of a certain divisor n.

(I)	1	2	3	4	5	6	...	?	...
(nI)	$1n$	$2n$	$3n$	$4n$	$5n$	$6n$	...	?	...

We can scan the second row (nI) until the dividend N appears. The required quotient is then the integer that is above this in the aligned first row (I). With suitable choices of n and N, it is always possible through this clear, but lengthy, procedure, to find the quotient of the <u>division</u> of N by n, which we write as $N \div n$ or $n \overline{)\, N}$.

The easiest way of reducing the length of the procedure outlined above would be to use multiples of n that may be already known; for example, $n \times 10$. Once students know that the challenge now is to <u>shorten</u> the procedure, they will suggest their own "partial quotients." They can then write down only those integers of (I) and (nI) which will lead them closer and closer to (N).

(I)	...	10	...	20	...	?
(nI)	...	$10 \times n$	...	$20 \times n$	...	N

But since they know N and n as given, they can save work by reducing N, subtracting from it the multiples of n they know and find easy to use. This awareness allows them to retain from (I) and (nI) only those integers that are "actual partial quotients;" corresponding to a part of N that we can subtract in advance. This may be illustrated by an example, say $896 \div 16$.

$$
\begin{array}{llllll}
\text{(I)} & 10 & + & 20 & + & 20 & & ? \\
\text{(16I)} & 160 & + & 320 & + & 320 & & 896
\end{array}
$$

Here the dividend has been approached through the multiples 16×10, and then 16×20 which has been used twice. Only the sum of these is recorded in (nI). We have $50 \times 16 \sim 800$ so that we now have only 96 to deal with in the dividend, and this can be found in a similar sort of way; so that, finally, we have $896 \div 16 \sim 50 + 5 + 1 \sim 56$.

$$
\begin{array}{ll}
\text{(I)} & 50 \\
\text{(16I)} & 800 \,|\, 896
\end{array}
$$

$$
\rightarrow \quad
\begin{array}{l}
5 \\
80 \,|\, 96
\end{array}
$$

$$
\rightarrow \quad
\begin{array}{l}
1 \\
16 \,|\, 16
\end{array}
$$

What is important here is that students use what they know and have mastered; that each step replaces one division by another in which the dividend is smaller and perhaps more easily managed; and that though there will be one answer, there will be various individual routes to it.

Recording in vertical notation may make the awareness sharper.

$$
\begin{array}{rl}
16\overline{)896} & \quad 10 \qquad \text{(partial} \\
-160 & \qquad\quad \text{quotients)} \\
\hline
736 & \quad 20 \\
-320 & \\
\hline
416 & \quad 20 \\
-320 & \\
\hline
96 & \quad\ 5 \\
-80 & \\
\hline
16 & \quad\ 1 \\
-16 & \\
\hline
0 &
\end{array}
$$

The appearance of the zero at the end of the division in vertical notation is a sign that the choice of n and N was such that there is an integer quotient q such that $n \times q \sim N$. Clearly, if the choice of dividend had been, say, one unit more, then the last subtraction would yield a 1. Any integer that does remain like this at the end is called the <u>remainder</u>, usually indicated with an r — thus, for $899 \div 16$ we get 56 again as a quotient but have a remainder 3, so we write $899 \div 16 \sim 56\ r3$.

In the traditional routine, teachers insist that students find "best" partial quotients by using the highest possible power of ten each time. In the example 896÷16, this first best partial quotient would be 50 — it is customary to record only the 5 in the appropriate place above the dividend, in this case above the 9 — and the next partial quotient would be 6.

$$
\begin{array}{r} \\ 16\overline{)896} \end{array}
\quad \rightarrow \quad
\begin{array}{r} 5 \\ 16\overline{)896} \\ \underline{800} \\ 96 \end{array}
\quad \rightarrow \quad
\begin{array}{r} 56 \\ 16\overline{)896} \\ \underline{800} \\ 96 \\ \underline{96} \\ 0 \end{array}
$$

Clearly this involves partial quotients and the same sequence of operations as before. But it is an unnecessary burden to the student to have to seek the "best" for the "first." Simply by relinquishing this demand, the so-called "long division" algorithm regains some common sense and can be mastered in a relatively short time — for some students in a very short time.

Summary

This chapter was written for mathematicians who already know it all, but may have never thought that a new foundation for their science could be based on language. It was also written for elementary school teachers who feel they are far from knowing it all, but who would like so much to help their young students; in particular, to help them to be in control of the four operations on integers, to make sense of the various algorithms, and to stop feeling frustrated and dependent to the point of paralysis.

The foundation on language that is proposed here could never have been considered preferable to others, until ways of involving students had been found that enable them to be quite clear that some things are given to them, but that it is mainly their responsibility to grow in awareness and in ability to perform all the operations correctly.

This is why it was necessary to produce a tightly knit curriculum, giving the indispensable details, but with obvious extensions left to the reader. This curriculum — summarized in the second headings — looks very different indeed from the traditional one. But it still meets the needs of elementary teachers, for nothing of what they are currently employed to teach has been neglected.

By starting with language, we take advantage of so much learning and so many ways of learning evidently at work in the students' ability to speak their native tongue. Revealing the reality of numeration as a linguistic challenge, we give students the chance to be in control of this chunk of their language and to reach a level of competence equal to that of any knowledgeable adult. In extending numeration to numerals of many digits — up to 15 — we force awareness that this can be achieved easily and at a very small cost in ogdens.

By extending numeration to various bases, we force awareness that there are conventions which can be changed easily. We do not need to work this topic out exhaustively. It seems important that young students become sensitive to the various ways of working of their minds. Hence, students are invited, from time to time, to do what seems accessible to them, even though this is not presented systematically to them at this stage. This applies not only to bases of numeration, but also to "place-value," to "transformation," and to "equivalence," which we know young children encounter as a matter of course in their use of language.

By placing "order structure" where we do, we emphasize that it is not necessary to focus on it in the study of numeration and that it leads us to "counting" in the oral sense — a linguistic awareness. The other sense of counting — "counting elements of a set" — is a much more complex activity and this justifies postponing it. This is a specific difference between our foundation of arithmetic and others.

The second aspect of counting adds a new dimension to numerals; we therefore now call them "cardinals." One development is "complementarity," which is easy to present, using the set of fingers of our hands. This gives us a remarkable opportunity of showing that complementarity is the foundation of addition and subtraction — each of these corresponding to a particular awareness of the subsets of fingers created by folding some of them.

Only when addition and subtraction become the source of new awarenesses, do we need to think of the elements of the sets involved as displaying a new property — that of being linked by being connected to a "unit." This means that new strings of equivalences can appear as "other ways of saying" what has been said before. At this stage, we have a "theory of number." Statements can be made which are true within a certain universe of experience, that of "integers;" for example, that each integer is equivalent to the class of its partitions. These enrich our awareness of "number" beyond their connection with numerals or cardinals.

Calculations with numbers involve transformations of the given, so that it is possible to say something about that given, making use of the dynamics of the mind called "equivalences." One of these transformations, called "finding the answer," is very important at school. But, of course, it is only one momentary part of the working of the mind — a passing moment, even if it is sometimes the last one.

Much that could be added at this stage is left out in order to introduce another pair of operations, multiplication and division. Here, we are guided by linguistic awareness. In another curriculum, published between 1957 and 1963, we forced a different awareness, which was as valid for teaching as the one now proposed, but not as rigorous mathematically.

The set of integers is looked at in a special way to yield new sets. The doubles, doubles of doubles, and so on, allow us to call the relationships involved "multiplications," and their inverse operators "reciprocals" — later to become divisors, and fractions. Only a very small part of this treatment is required to take care of the multiplicative links stressed at school under the name of "multiplication tables."

Our approach also generates parts of school algebra that are usually postponed several years. Algebra is another name for the awareness of the dynamics involved in the transformations of mental structures available to us all. Hence, this curriculum does not postpone what can become new awarenesses just because it has not been done before in this way. Linear equations using the four operations find a rightful place at this stage, some years before it is traditionally introduced.

Another distinguishing feature of the curriculum developed in this chapter is its central emphasis on the distributive law — and its inverse, factoring — which is required to ensure that multiplication is encountered in all its complexity, that is, as affected by the "polynomial" form of numbers. The classical algorithm of multiplication can then be derived easily and without mystery. The treatment of "long division" also makes this often dreaded operation friendly, allowing the exercise of imagination and caution.

The microcomputer course Visible and Tangible Mathematics contains a similar treatment of the material of this chapter, but has the advantage of showing how the use of the computer forces awareness without a word being uttered, and leaves students to decide how much practice they need individually to master any topic. (Further details are given in the appendix.) *

---

* See Appendix for descriptions of **Student Sheets** designed by author to introduce many of the mathematical awarenesses of this chapter to students.

11 Beyond Integers: Algebra and Language

Introduction

In the previous chapter, we transformed numerals into cardinals and then into the numbers called integers. In some of the exercises with the operation of multiplication, we found that the "reciprocal" of an integer could easily become part of the language available to beginners — with a clear meaning, that of a reverse reading of a relationship between integers.

Reciprocals seemed to appear naturally in these contexts and there was no reason to delay introducing them. As long as there was plenty to do with these "operators" as such, we could postpone looking more carefully at the notation to see what new awarenesses it may suggest. But now we can ask, for example, whether the 1 in the sign $\frac{1}{n}$ could be replaced by some other integer.

The notations used to describe the relationship between, for example, the sequences (I) and (2I) are recapitulated below.

- From (I) to (2I), the operator is "multiply by two," written ×2; for any element n in (I), the corresponding element in (2I) is "twice n," or "two times," written $2n$. Thus, there are two equivalent notations: $n{\times}2 \sim 2n$.

- From (2I) to (I), the operator is "multiply by half," written $\times\frac{1}{2}$; for any element $2n$ in (2I), the corresponding element in (I) is n. Thus, $2n \times \frac{1}{2} \sim n$.

  But note that we can also see the corresponding element in (I) as "half of" $2n$; and we shall later replace the "of" with ×.

The notation extends in the same way to other multiples and reciprocals. Note that, in general, multiples are named as so many times — as in "six times" — apart from the irregular forms "twice" (×2) and "thrice" (×3). Furthermore, reciprocals are generally named by attaching the suffix "th" to the name of the corresponding integer — as in "sixth of" — apart from the irregular forms "half of" ($\times\frac{1}{2}$), "third of" ($\times\frac{1}{3}$) and "quarter of" ($\times\frac{1}{4}$).

It is part of the work we can do with these operators to find that there is an algebra allowing the substitution of two or more successive ones by a single one. For example, $\times 2 \ \times 2$ can be replaced by $\times 4$ and $\times \frac{1}{2} \ \times \frac{1}{2}$ can be replaced by $\times \frac{1}{4}$ — as we did in the formation of (4I); in the same way, both $\times \frac{1}{2} \ \times \frac{1}{5}$ and $\times \frac{1}{5}$ $\times \frac{1}{2}$ may be replaced by $\times \frac{1}{10}$, and so on. Conversely, we may transform a given operation into a string of operations acting successively. Moreover, we could treat mixed strings of multiplications and reciprocals in the same way. For example $\times 4$ $\times \frac{1}{2}$ can be replaced by $\times 2$, and $\times 2 \ \times \frac{1}{4}$ can be replaced by $\times \frac{1}{2}$. These equivalences suggest the possibility of rewriting $\times 2$ and $\times \frac{1}{2}$ with the new notations $\times \frac{4}{2}$ and $\times \frac{2}{4}$. Thus $\times \frac{2}{4}$ acts on elements of (2I) to yield corresponding elements of (I) and $\times \frac{4}{2}$ acts in the reverse direction.

(I)   1   2   3   ...   n   ...

$$\uparrow \times \frac{2}{4} \quad \downarrow \times \frac{4}{2}$$

(2I)   1   4   6   ...   2n   ...

In general, then, we replace 2 by $\frac{2n}{n}$ and $\frac{1}{2}$ by $\frac{n}{2n}$, for any n.

This means that the expressions 2 and $\frac{1}{2}$ can be represented by infinite sets of pairs of integers. We can force this new awareness in the following equivalences — which are written in a "vertical" form.

$$2 \quad \frac{2}{1} \quad \frac{4}{2} \quad \frac{6}{3} \quad \ldots \quad \sim \quad \frac{2n}{n} \quad \ldots$$

$$\frac{1}{2} \quad \frac{2}{4} \quad \frac{3}{6} \quad \ldots \quad \sim \quad \frac{n}{2n} \quad \ldots$$

All the pairs of integers in each row are interchangeable "names" and can all equally well be used when naming operators; for example, $\times 2 \sim \times \frac{6}{3}$. There will be similar equivalences for 3 and $\frac{1}{3}$, and so on. In each case, the first row introduces another way of looking at an integer, namely as a *class of equivalent ordered pairs*. To keep this in mind we use a new name for the class itself, calling it a <u>rational number</u>. This adds a deeper insight to our development of entities which are written with one sign but carry different awarenesses.

Furthermore, the new notation forces a new awareness: since the second row above is also a class of equivalent ordered pairs we may also call it a rational number. Thus, the signs $\frac{1}{2}, \frac{1}{3}, \frac{1}{4}$, and so on, which have not as yet stood on their own, are now also taken to name rational numbers.

Once accustomed to this new notion, we can drop the word "rational" and just speak of numbers, until such time as another awareness may require a further qualifier for a while. For example, at the end of this chapter, some numbers will be further distinguished as not being rational, and so "irrational." Note that the names traditionally used here mean "with ratio" or "without ratio" and do not refer to the everyday meanings, "reasonable" or "unreasonable."

In some contexts the sets of ordered pairs of integers are also called <u>fractions</u>. Although we will still use this word to specify the kind of number we are concerned with, when we do so we shall <u>not</u> be referring to the etymology which emphasizes that these numbers first appeared when a whole and a fragmented part of it were considered simultaneously. So much of the confusion in the study of fractions, found in elementary schools, or even later, can be traced to the difficulty of making "a bit of something" function as an operator.

Algebra as Operations Upon Operations

In the previous chapter, we found that addition and subtraction can be used together, as well as separately, in the solution of some arithmetical problems. We also found that multiplication could merge with addition, subtraction and reciprocation to produce "linear equations." In every case, two operations were replaced by one, and conversely; we operated on operations.

This aspect was present in the minds of the founders of that branch of mathematics historically called algebra; but it was rarely used consciously. The Arabic root of the word refers to a jump of some number from one side of a relationship to the other. Nothing of this intuition remains in our presentation. In the science of education, which is founded on awareness, we take algebra to refer to a mental activity, rather than to its strictly historical and mathematical content.

Human intellects manipulate many notions and can generate new aspects which, when singled out, become other, possibly new, notions. If, instead of locking at these aspects, we become aware of the dynamics behind them, then we stress something present in the mind, that can be felt as the dynamics sustaining the mental process. We say we become aware of the algebra, or the algebras, making up these mental dynamics.

Hence, for us, <u>algebra</u> is another way of speaking of the mental dynamics necessary to transform some mental given into another mental form, which is kept related to the first. Algebra will appear in mathematics — which is a mental activity — alongside other mental activities involving numbers, images, sets, or complex entities perceived as wholes called structures. Algebra, as the dynamics of the mind, then becomes — and can be seen in — specific things like equations, polynomials, mappings and so on. These new realities, which are easily distinguishable and capable of being relate to each other in ways created by the minds entertaining them, are called algebraic entities.

There is great advantage in extending the meaning of the word "algebra" so as to be able to stress that it is concerned mainly with mental dynamics, rather than only with mathematical operations. It places mathematics back in its proper context, which is mental and concerned with relationships of awarenesses. On the other hand, it allows us to find algebra wherever it is; for example, in the processes leading to the generation of language, or to the mastery of reading, which do not seem <u>a priori</u> to have much to do with algebra in its traditional sense.

But, more than anything else, it turns newcomers into mathematicians. This is a new notion in education and needs a little elaboration. Mathematicians are specialists; they sometimes understand very little of what other mathematicians are talking about. However, it would not follow that they cannot do good work in their own particular area of interest and activity. If we agree that this is true for everyone, we would be able to look at their ability to work, and the quality of their involvement, in a particular area, and judge whether each person is independent, autonomous and responsible in that area.

From this point of view, mathematics teachers can say whether such and such a student is as good as they are in a particular field. Since working through awareness leads to mastery at every step, however small that is, teachers can say again and again: "In this, you are as good as I am."

In fact, our concern will bring learners face to face with their own activity, reducing memorization to the minimum — into what cannot be invented, giving as much practice as each student needs, and enabling a thorough survey of an area of work about which students will speak with aplomb and only of what they know and are sure of knowing. For example, in our study of addition we have forced the awareness that when a problem is given, its form can be altered to become one from which the answer can be simply read.

The transformations that do this are mental operations, as they are for mathematicians. These mental operations include more than the traditional four or arithmetic — addition, subtraction, multiplication, and division. We have already come across iteration, which leads to exponentiation, and reciprocation or reversal, which leads to fractions. Furthermore, we have seen how the equivalent expressions of everyday language lead to "equivalence," and so to the "classes of equivalence," which are the backbone of calculation. When equivalence spontaneously springs to mind, this allows us to say that algebra is indeed another name for mental dynamics.

Made sensitive to equivalence by the use of language, all students will feel comfortable with our insistence that the entities we meet in our mathematics classes are only "single" elements for our perception. These also trigger the equivalence classes that include them, so that the proper form required by a particular challenge can be easily retained. For example, $\frac{1}{2}$ is one thing when seen as marks on paper, but another when seen as a family of equivalences, any one of whose forms is readily available for use. This means that we offer our students an education of their awareness, in which shifts of focus are part of a very natural way of working. The given is not sacrosanct, it is only a trigger for other, possibly more appropriate, forms. This is precisely what mathematicians do; and what our students could do from the start.

Once it is understood that algebra is the combination, the merging, of operations — in other words, operations upon operations — then the approach to students faced with some specific mathematical challenge seems charted for us, their teachers.

Adding and Subtracting Fractions

As an example of such an approach, we consider briefly the addition or subtraction of fractions. Two given fractions are represented by pairs $\frac{a}{b}$ and $\frac{c}{d}$. Our perception sees only these two forms, but our minds can be made to evoke at once the two equivalence classes to which they respectively belong. With this wealth in mind, we tell our students that in order to do the addition or subtraction we need to look into the meaning of these operations.

Everyone using language properly will <u>know</u> that there is no meaning in adding objects which are labeled in a way that distinguishes them. For example, apples and pears may be lumped together as fruit by a mental operation, but this requires us to ignore that which distinguishes them and to conceive of them both as fruit. Language is adequate for such assimilation into a broader class than those established by each separate component.

Without such a mental operation, we cannot begin to consider adding, which is based on continuing counting elements from one set to those of the second. It is language that requires that we first find "a common name" for the elements of the two sets before we can proceed. So it is language, not arithmetic, which tells us that if we want to add $\frac{a}{b}$ and $\frac{c}{d}$ it will be necessary first to find a common name — there may be more than one — and then replace the given forms by the equivalent expressions which make addition possible. It is the addition that forces this, not the fractions as such.

We can give fractions names by recalling how they arose from mergings of operations. For example, $\times 2 \times \frac{1}{3}$ can be replaced by $\times \frac{2}{3}$, and the form $\frac{2}{3}$ may be seen as "twice," or "two of," whatever it is that is represented by $\frac{1}{3}$. So the second part of the ordered pair — below the bar in "vertical" form — names the object being considered; we call it the <u>denominator</u>. The first part — above the bar — then "enumerates" how many of the objects named by the second part are being taken; it is therefore called the <u>numerator</u>.

Once they are named, fractions may be added or subtracted. The first step is to look at the equivalence class of each given form.

$$\frac{a}{b} \sim \frac{2a}{2b} \sim \frac{3a}{3b} \quad \cdots \qquad \frac{Na}{Nb} \quad \cdots$$

$$\frac{c}{d} \sim \frac{2c}{2d} \sim \frac{3c}{3d} \qquad \cdots \frac{Nc}{Nd} \quad \cdots$$

The next step is to look for a multiple of b which has the same numeral as a multiple of d. Note that we invoke numerals here, rather than numbers, in keeping with the linguistic awareness that is guiding us, although number would be more correct in the context of operations. We may find many such pairs of multiples; but <u>one</u> pair is certain to be noticed, namely $b{\times}d$ and $d{\times}b$. The initial pair of fractions may now be replaced by

$\frac{a \times b}{b \times a}$ and $\frac{c \times b}{d \times b}$. Since $b \times d$ (or $d \times b$) is a <u>common</u> denominator the two forms can now be added as required — subtraction may be treated in a similar way.

$$\frac{a}{b} + \frac{c}{d} \sim \frac{a \times d}{b \times d} + \frac{c \times b}{d \times b} \sim \frac{a \times d + c \times d}{b \times d}$$

Note that working with letters instead of numbers excludes actual numerical computation, so that we can just stress the operations. Thus, we can stay with the algebra of the situation. The letters prevent the actualization of the addition and force awareness of <u>what is being done</u>. This is algebra, since it stresses operations.

All this shows how to educate awareness in the case of addition — or subtraction — of fractions, or how to use sane awarenesses to know what has to be done and why. Later this may be given a more formal appearance to look like an axiomatic system. Mathematicians know all the above steps in detail and prefer a logical presentation; but here, it is more appropriate to put the stress on awareness and mastery, as opposed to acceptance and memorization.

Clearly, the fraction equivalent to the sum or the difference of two fractions is also an equivalence class with an infinite number of forms. It is then a <u>new</u> problem, to find which of the elements of this class shall be retained and called <u>the</u> answer to the given operation. it is sometimes considered "wrong" to take any member of the class as the answer. An additional step, called "simplification" of the fraction may then be required; this consists in finding a form where a and b are not both multiples of some integer N other than 1 — in which case they are said to have no "common factor;" for example, $\frac{6}{8}$ can be "simplified" by replacing it with the equivalent $\frac{3}{4}$.

An oscillation is required to replace a perceived item by its equivalence class. With this goes its reversal, where we see a class as represented by anyone of its members, more particularly the so-called simplified form. This is all part of the behavior as mathematicians that we can, and perhaps should, give our students.

This section has indeed taken us beyond integers and shown why it is of greater importance in our students' education to force awareness of algebra as operations upon operations, and to remain in contact with language and with the dynamics of the mind, if we want to do the work as well as young children do in their spontaneous learnings: that is, do things once and well and forever.

Operating with Fractions

The sequences of integers (nI) provided a natural way of introducing fractions as operators. They appeared first as "reversals of multiples," or "reciprocals," and they then became entities in their own right when we found in a new examination of how these operators were generated that, like integers, they formed classes of equivalence. The notation for reciprocals — $\frac{1}{n}$ — was handy and could be extensively used to express some of the relations between the sequences of multiples. The extension of the notation to the form $\frac{a}{b}$, where a and b are both integers, arose through the merging of two operators to establish a single relation between two sequences; for example $\times 2 \times 1\frac{1}{3}$ could be replaced by $\times \frac{2}{3}$ which operates on, say, elements of (3I) to give corresponding elements of (2I).

Since a fraction has infinitely many equivalent forms, then it is, above all, an <u>infinite equivalence class</u>. This important notion has already arisen in the case of integers where these were seen to be equivalent to infinitely many subtractions. An integer is, of course, also an equivalence class of additions, but this is a finite class. But now an integer can be represented by another infinite equivalence class — arising from the reciprocal that corresponds to the integer; for example, the following equivalences for 1:

$$1 \sim \frac{1}{1} \sim \frac{2}{2} \sim \frac{3}{3} \qquad \ldots \sim \frac{n}{n} \ldots$$

Multiplication by an integer followed by multiplication by its reciprocal — or conversely — is equivalent to an "identity" relation — an operator which keeps things as they were; for example, $\times 2 \times \frac{1}{2} \sim \times 1$. This means that the operator relating an integer to itself, as in the pairs $\frac{n}{n}$, can be read in two ways, generating new awarenesses. Thus, $\frac{n}{n}$ may be read as equivalent to $n \times \frac{1}{n}$, which is a multiple of, or "n of," $\frac{1}{n}$. But it may also be read as equivalent to $\frac{1}{n} \times n$, which is then read as "one n-th of" n. Note that the $\times$ signs are here read as "of." Furthermore, we may also read $\frac{n}{n}$ as "n divided by n," a purely linguistic convention that can be written in the form we have already used, $n \div n$.

These forms can all be extended to general fractions $\frac{a}{b}$, where a and b are any integers. Thus $\frac{a}{b}$ may be read "a of one—bth," "one b-th of a," or "a divided by b," and written in the following forms.

$$\frac{a}{b} \sim a \times \frac{1}{b} \sim \frac{1}{b} \times a \sim a \div b$$

Here, we are extending our language. Each new verbal expression corresponds to a new awareness — not necessarily a new notion, but a different understanding of what has already been met. In this way, we know we remain in contact with the mental powers which enabled the remarkable evolution of language in early childhood. We can expect that becoming a mathematician in these circumstances will be facilitated — but also with vastly improved results. This may mean doing in one year at school work which now requires several years, and doing it so much better.

* * *

We have not presented multiplication as repeated addition, though in a way it is such an iteration. One advantage of not doing this is that we can still "multiply by one" which is certainly not a repeated addition. Multiplying by one is an "identity" operation, leaving things as they were. This is a very different awareness from an operator which generates change.

In the previous chapter, we also met another identity operation when we made "zero" into a number as well as being a place-holder used with numerals. Adding or subtracting zero extends the notions of these operations by leaving things unchanged, but still <u>imagining</u> that an addition or subtraction has taken place. The "nothing" of everyday experience was given a new name, zero, and written 0, so that we could write the equivalence $a + 0$ ~ a.

Leaving things unchanged can also be considered an iterable operation. This takes but the time to think it, but it produces no perceptible effect. In other words, expressions like $a+0$ and $a\times1$ have some reality for us, for we can entertain the notion of <u>no change</u>; though perhaps not so easily as <u>change</u>, which has marked our life from the beginning — because living takes place in time, and time flows and can be perceived as flowing.

Furthermore, multiplication has not been perceived as a "magnifier," since it was applied from the start to reciprocals which, in fact, do the opposite. Repeated addition <u>is</u> a magnifier and students sometimes get confused when they are asked to multiply and end up with something which is smaller than what they started with. Similarly, division as a repeated subtraction seems to suggest diminution, yet dividing by a fraction can be a magnification; for example, in $1\div\frac{1}{2}$ ~ 2. Our approach avoids these confusions while preserving recognized mathematical content.

What has been presented is that there is always more to be read in a relationship than appears at first sight. For example, the equivalence 2×3~6 can suggest many other readings.

- 2×3~6 or 3×2~6 (expressing that multiplication is commutative);

- 6~2×3 or 6~3×2 (reading the relationship the other way round — since, in technical language, equivalence is a symmetric relationship);

- $3 \sim \frac{1}{2} \times 6$ or $2 \sim \frac{1}{3} \times 6$ (by reversal of ×2, ×3 — namely, "inverting" multiplication) and so also further readings of these by symmetry and commutativity;

- 6÷2~3 or 6÷3~2 (the notation of <u>division</u> representing a new awareness — that $\times \frac{1}{2}$ or $\times \frac{1}{3}$ may be read as "dividing by 2" or "dividing by 3," respectively).

If we make sure that our students see <u>all</u> these facts of awareness every time they look at any <u>one</u> of them, we have again made mathematicians out of them as far as some area of the school curriculum is concerned.

* * *

Multiplication of a fraction by a fraction is better understood when we ask: what fraction is equivalent to a fraction of a fraction? This leads to a new equivalence class for fractions. We already know that $\frac{1}{n} \times n$ is an identity operation. We also know that every fraction has an infinite number of equivalent expressions. By bringing these two awarenesses together, we have the following sequence of equivalences, in the writing of which some multiplications are abbreviated into juxtapositions — for example ac for $a{\times}b$.

$$\frac{a}{b} \times \frac{c}{d} \sim \frac{ac}{bc} \times \frac{cb}{db}$$

$$\sim ac \times \frac{1}{bc} \times cb \times \frac{1}{db}$$

$$\sim ac \times (\frac{1}{bc} \times cb) \times \frac{1}{db}$$

$$\sim ac \times \frac{1}{db}$$

$$\frac{a}{b} \times \frac{c}{d} \sim \frac{a \times c}{d \times b}$$

Here, the $\times$ has different meanings on each side of the equivalence: on the left it represents "of," on the right it represents multiplication of integers. Once it is understood how these replacements are made, there is no danger in changing the language describing them. Thus, we may now say that the "product" of two fractions is another fraction whose numerator is the product of the numerators, and whose denominator is the product of the denominators. This means that × may now represent a multiplication of fractions as well as of integers.

A "chain" of fractions, linked by "of," may be worked out step by step, with successive pairs of fractions being merged into one fraction in the way that has just been described. For example, "half of a half of a half" may be expressed as $\frac{1}{2} \times \frac{1}{2} \times \frac{1}{2}$, which is equivalent to $\frac{1}{2} \times \frac{1}{4} \sim \frac{1}{8}$. By reversing the process we can say that any fraction is equivalent to an infinite number of chains of fractions.

$$\frac{m}{n} \sim \frac{m}{a} \times \frac{a}{n}$$

$$\sim \frac{m}{a} \times \frac{a}{b} \times \frac{b}{n}$$

$$\sim \frac{m}{a} \times \frac{a}{b} \times \frac{b}{c} \times \frac{c}{n} \cdots$$

Such chains of 2, 3, 4, . . . fractions are generated through the identity operation by ensuring that when an integer is inserted as a denominator it is also inserted as a numerator, so as to "cancel" the effect of the first insertion.

The different words "product," "multiplication," "of," refer to awarenesses which can be linked together mentally in the way that words can be linked in everyday language. We use one or the other word according to the particular need. This power makes <u>equivalence</u> essential in mathematics. It allows the mental activity of "being with" at the same time as it allows shifts of consciousness. To transfer this notion from language, to place it under the strong light of awareness, and to find that it can work in a new context, is part of the awareness, and to find that it can work in a new context, is part of the awareness of the dynamics of relationships — that is, of mathematics.

To educate the mathematician in all students is to force such awarenesses on them. This will then take care of curricular contents, which are simply lists of interrelated, topics generated by special awarenesses. Hence, there are great advantages in concentrating on mental activities which deliberately embrace many awarenesses and which can lead to a new one. When such a new awareness is stated in words — or notations — we can see it as a <u>by-product</u> of these activities. Such by-products are traditionally seen as the main products, whereas for us, the activities capable of educating the students, are indeed the product — the product of education.

Readers can now see that in the science of education the approach used to make mathematicians out of students has to put the stress on mental activities, yielding by-products which form the knowledge of mathematics sought by all teachers, but not always seen as such by their students.

General statements, like the ones above, are not necessarily the outcome of any special virtue in the multiplication of fractions. They could have been suggested by other topics; but fractions happened to be the context in which they have been stated here. Awareness can be educated; and we try to show how, whenever possible.

Dividing Fractions

Another example, illustrating the importance of our emphasis on language in forcing mathematical awareness, is the division of fractions, which will be presented here independently of multiplication of fractions.

The first awareness is that the equivalence $2 \times \frac{1}{2} \sim 1$ can be read "there are two halves in one." This answers the question, "how many halves in one?" This is written — in a vertical notation — as follows:

$$\frac{1)\overline{1}}{2} \quad \rightarrow \quad \frac{1)\overline{1}^{\,2}}{2}$$

We can now extract as much as we can from this awareness. In the first place, we can now ask how many halves there are in two — or in three, in seventeen, in n; thus, there are 4 halves in 2, $2n$ halves in n. We could then ask how many thirds in 1; there will be 3 since $3 \times \frac{1}{3} \sim \frac{1}{3} \times 3 \sim 1$, and this can be extended to finding how many thirds in other integers. Finally, we can extend this to the reciprocals of any integer.

In general, we may ask how many $\frac{1}{a}$ "go into" b, the answer being $a \times b$, or ab. This awareness is not a "formula" to remember, but a perception of the challenge and how one can answer it at once, because one has understood the algebra behind the question "how many halves in one?" If the question is put in other forms — for example, "b divided by $\frac{1}{a}$," "divide b by $\frac{1}{a}$" or "$b \div \frac{1}{a}$" — it is then part of the new awareness that these are also dealt with more easily, and with understanding, when transformed into the "goes into" form.

* * *

The division of fractions needs two further awarenesses that need to be worked on. First we consider what happens when the b in $b \div \frac{1}{a}$ is replaced by a fraction, say $\frac{m}{n}$. Since $\frac{5}{7}$, to take a specific example, is also $5 \times \frac{1}{7}$, and the reciprocal is read with the suffix "th" as part of its name, then $\frac{5}{7}$ is read "five sevenths." Similarly, $\frac{m}{n}$ is read "*m n*-ths." If we ask "how many $\frac{1}{a}$ go into $\frac{m}{n}$?" with a loud emphasis on the numerator and with a whisper for the denominator, then we hear "how many $\frac{1}{a}$ to into ***m*** *n*-ths?" — and the answer is "*ma n*-ths." This can be written $\frac{ma}{n}$, and in two further equivalent ways corresponding to the forms "go into" and "divided by."

$$\frac{ma}{n} \quad \sim \quad \frac{1}{a} \overline{\smash{\big)}\,\frac{m}{n}} \quad \sim \quad \frac{m}{n} \div \frac{1}{a}$$

A second awareness now appears when we also replace the $\frac{1}{a}$ by a general fraction. This awareness can teach us a great deal about the way the mind has to be used in order to remain with a question. So it needs spelling out more explicitly in this context.

96

We start with another specific example: "how many $\frac{2}{3}$ go into

1?" We know that $3 \times \frac{1}{3} \sim 1$; moreover, since 1 is half of 2, we also

know that $\frac{1}{3}$ is half of $\frac{2}{3}$. If there are three thirds in one, there

are <u>also</u> "one third and two thirds," the "and" here can be read,

at will, either as conjunction or as addition. So, $\frac{1}{3} + \frac{2}{3} \sim \frac{3}{3} \sim 1$.

There is indeed <u>one</u> $\frac{2}{3}$ in 1, leaving a remainder of $\frac{1}{3}$. If, as

shown above, we read this remainder as half of the $\frac{2}{3}$, then we

can see that $\frac{2}{3}$ goes into 1 "one and one half times."

By working through a certain number of specific examples to ensure awareness of the shift from one awareness to the other, we can see that we have solved the problem of dividing fractions without having had to "fracture" anything in order to generate the fractions. Awareness of language is sufficient to meet all the obstacles and subtleties which arise in this context.

Division of fractions can be expressed in two ways — "how many

$\frac{a}{b}$ in $\frac{c}{d}$?" or "$\frac{c}{d}$ divided by $\frac{a}{b}$?" — but both yield the same

solution, $\frac{cb}{ad}$. Thus, there is always <u>one</u> fraction equivalent to "one fraction divided by another," and its equivalence class can be used to give it as many forms as we want.

Traditionally, the division of fractions is linked to multiplication. This takes for granted that each operation is the inverse of the other in the case of fractions as well as integers. But what we find, when we keep the operations distinct, is that the "product" of one fraction by the reciprocal of another is equivalent to the "quotient" of the first by the second, <u>and</u>, conversely, the quotient of two fractions is equivalent to the product of the first by the reciprocal of the second.

Rational Numbers

The development of the four operations on fractions makes it clear that we have to work with <u>rational numbers</u>, namely equivalence classes, and that we have to shift our awareness from one focus to another one, relating these in our minds.

We can <u>unify</u> the field of fractions and integers, by showing that integers can be considered as fractions with unit denominators: any integer is equivalent to $\frac{a}{1}$. This is a new awareness and a new notation — apart from the case $\frac{1}{1}$ which was part of the treatment of $\frac{n}{n}$. Since $\frac{a}{1}$, as a fraction, can also be written $a \times \frac{1}{1}$

and $\frac{1}{1}$ can be written as 1, then we have the following equivalences.

$$\frac{a}{1} \sim a \times \frac{1}{1} \sim a \times 1 \sim 1 \times a \sim a$$

The four operations can be combined and merged according to the algebras of the two pairs, addition and subtraction, multiplication and division. A cascade, like that given below, will merely take longer to "simplify" when the letters stand for fractions rather than integers: the algebra will be the same, though the writing out of successive equivalences will be different.

$$\frac{A+B}{C} + D \times (E \div F) \times \left(\frac{G}{H} - \frac{J+K}{L+M} \right)$$

Finally, to complete the unification of fractions and integers, we need to consider the ordering of fractions. To find which of $\frac{a}{b}$ and $\frac{c}{d}$ is the largest, we select two fractions having the same denominator from each of the two equivalence classes. The fractions may now be ordered by their numerators. Thus, we use language to let us choose between linguistically comparable forms. For example, if the common denominator is taken to be *bd,* then the two numerators are *ad* and *cb,* and these may be ordered as integers.

* * *

So far, fractions have been either operators or ordered pairs of integers. These correspond to two linguistic awarenesses that have different mathematical connotations. An <u>operator</u> is a true algebraic entity which allows us to <u>do</u> something, such as generate one number out of another — for example, $\frac{1}{5}$ of 10, or $\frac{2}{3}$ of 24. An <u>ordered pair</u> gives us an equivalence class which may be required in other circumstances; for example, to add $\frac{1}{2}$ and $\frac{1}{3}$ needs the selection of equivalent fractions with the same denominator. The mathematician in every student will know which of these aspects are required in a particular context.

When we write $\frac{a}{b}$ for a fraction, what *a* and *b* are is left open. The order structure on the integers gives us three possibilities: *a* is less than, equal to, or greater than, *b*. Sometimes, it is only in the first case, *a*<*b*, that $\frac{a}{b}$ is called a fraction — retaining the notion of a fraction as "a bit of something." But here we allow all cases. Thus, we call $\frac{a}{a}$ a fraction, for this is the form of an "identity operator." Moreover we also include the third case, $a>b$, so that $\frac{6}{5}$, say, is as much a fraction as $\frac{5}{6}$. In this case we can write $\frac{6}{5}$ as $\frac{(5+1)}{5}$, this being equivalent to $\frac{1}{(1+5)}$, called a "mixed number? and traditionally re-written by juxtaposing the integer and the fraction parts, $1\frac{1}{5}$ .This form is but a relic of the notion of a fraction as part of a whole, a historical aspect which does not need to be retained.

Decimals: A New Language

It occurred to someone, centuries ago, to propose that we read and write fractions with denominator 10 in a new way. This proposal has been adopted universally, even though not with exactly the same words and notation everywhere. So we can give it to young students, in the first place, as an exercise in translation. "Decimals" are not <u>new</u> fractions, but they do have attributes which make them more useful in some contexts. It is a mark of the human side of a mathematician to choose any way of saving time and effort, and decimals do that.

The new language is first learned through a set of exercises in which we concentrate on the words and the notation. We replace the fraction $\frac{1}{10}$ by the form ".1", which is read "point one."

Similarly, $\frac{2}{10}$ ~ .2, which is read "point two," and so on for similar fractions up to $\frac{9}{10}$. Fractions with denominator 10 but with numerators greater than 9 are expressed in "mixed" form. For example, $\frac{11}{10}$ ~ 1 + $\frac{1}{10}$ ~ 1 + .1, which is usually written in the condensed form 1.1, read "one point one;" similarly, $\frac{115}{10}$ 11.5, $\frac{110}{10}$ ~ 11.0, $\frac{105}{10}$ ~ 10.5, and so on. In keeping with this form, fractions with numerators less than 10 are sometimes written with a zero to the left of the point — for example, $\frac{9}{10}$ ~ .9 ~ 0.9.

Once such translations and their converses have become second nature, the adding and subtracting of these new decimals is relatively simple; some examples are given below.

$$.3 \ + \ .2 \ \sim \ \frac{3}{10} \ + \ \frac{2}{10} \ \sim \ \frac{5}{10} \ \sim \ .5$$

$$.7 \ + \ 3.2 \ \sim \ \frac{7}{10} \ + \ \frac{32}{10} \ \sim \ \frac{39}{10} \ \sim \ 3.9$$

$$.5 \ - \ .3 \ \sim \ \frac{5}{10} \ - \ \frac{3}{10} \ \sim \ \frac{2}{10} \ \sim \ .2$$

$$1.5 \ - \ .9 \ \sim \ \frac{15}{10} \ - \ \frac{9}{10} \ \sim \ \frac{6}{10} \ \sim \ .6$$

Such examples force awareness that when addition or subtraction is written in vertical notation the decimals need to be written with their points aligned. In this case the operations may be carried out exactly as they were with integers.

$$
\begin{array}{cccc}
.3 & .7 & .5 & 1.5 \\
+.2 & +3.2 & -.3 & -.9 \\
\hline
.5 & 3.9 & .2 & .6
\end{array}
$$

Hundredths, thousands, and all other fractions having powers of 10 as denominators, can be treated in the same way, once the corresponding new names and notation are given. Thus, $\frac{1}{100}$ is written .01 and read "point zero one," and $\frac{1}{1000}$ is written .001 and read "point zero zero one." These can also be written in the alternative forms, 0.01 and 0.001, which may be helpful in practice, since they retain the same number of zeros in each case.

When aligning decimals for addition or subtraction, it is sometimes helpful to rewrite decimals by placing any number of zeros to the right of the last digit, or to the left of the first digit. For example, to add 2.4, .007, 14.015, and 9.9, it helps to rewrite the decimals in the following vertical form.

$$02.400$$
$$+00.007$$
$$+14.015$$
$$+09.900$$

* * *

The multiplication of decimals is derived from the multiplication of the corresponding fractions, whose denominators are powers of 10. Consider the following specific example.

$$19.025 \times .06 \sim \frac{19025}{1000} \times \frac{6}{100}$$
$$\sim 19025 \times \frac{6}{1000} \times 100$$
$$\sim \frac{114150}{100000}$$
$$\sim 1.1415$$

The calculating procedure clearly can be condensed. First, replace the multiplication of decimals by a multiplication of integers, found by omitting points. In the given example, the second decimal yields 06 which is written in the simpler form 6, so the required product is 19025×6 ~ 114150. Now insert a point in the product in such a way that the number of digits to the right of the point is the total number of digits to the right of the points in the two original decimals. In this example, the point is inserted so that there are 3+2 ~ 5 digits to the right of it.

For division, we multiply one decimal by the reciprocal of the other. This means that the calculating procedure would be to first divide the numerators of the corresponding fractions as a normal division of integers. Then, insert a point in such a way that the number of digits to the right of it is now the <u>difference</u> of the number of digits to the right of the points in the two original decimals. For example, to divide 19.025 by .05, first divide 19025 by 5 giving a quotient 3805, then insert the point so that there are 3-2 ~ 1 digits to the right of it, so giving 380.5

We have not considered "exponentials" in this, or the previous, chapter, nor have we introduced "negative powers." This means that we have to leave the description of the above calculating procedures at a certain level of awareness. There are not real difficulties in extending these awarenesses to new notations and this has been done in the series of textbooks, Mathematics with Numbers in Color. But in this context, we are only sketching a curriculum. Rather than showing how every part of elementary mathematics may be treated, we prefer to concentrate on the central themes: *educating mathematical awareness* and *making students act as mathematicians.*

* * *

So far, only certain fractions have been written as decimals. We already know from previous work that there will be various other fractions that will be equivalent to fractions with denominators that are powers of 10. For example, fractions with denominators that are powers of 2, or powers of 5, or products of powers of 2 and powers of 5, may all be replaced by fractions with denominators that are powers of 10.

$$\frac{a}{2} \times 2 \times 2 \sim \frac{a}{8} \sim a \times \frac{125}{1000}$$

$$\frac{a}{5} \times 5 \sim \frac{a}{25} \sim a \times \frac{4}{100}$$

$$\frac{a}{2} \times 2 \times 5 \sim \frac{a}{20} \sim a \times \frac{5}{100}$$

But what about all the other fractions, those with denominators that have other factors than 2 or 5? This challenge is of interest to us because it leads easily to important new mathematical awarenesses.

To begin with, we note that the notation $\dfrac{a}{10}$ can be extended to include the case where a is a fraction. For example, when a is $\dfrac{20}{3}$, we can write $\dfrac{a}{10}$ in vertical form, reading this as "twenty-thirds over ten." This means that we can now replace any fraction by a form with denominator 10. For example, $\dfrac{2}{3}$ may be replaced by the following equivalent forms.

$$\frac{10 \times 2/3}{10} \sim \frac{20/3}{10} \sim \frac{6+2/3}{10} \sim \frac{6}{10}+\frac{2/3}{10} \sim .6+\frac{2/3}{10}$$

The last term in the above example may now be rewritten as an equivalent fraction with denominator 100, its numerator then being $\dfrac{20}{3}$. This fraction can then be treated in a similar way.

$$\frac{2/3}{10} \sim \frac{20/3}{100} \sim \frac{6+2/3}{100} \sim .06+\frac{20/3}{1000}$$

The process may be repeated again and again as long as we wish; the resulting terms being added together to yield a decimal form for the original fraction.

$$\frac{2}{3} \sim .6+.06+.006+\ldots \sim .666\ldots$$

The three dots used in the first equivalence indicate that we can go on adding further decimals with a number of zeros to the right of the point followed by a 6; and in the second that we can go on writing 6's on the right. This may also be indicated by writing just one 6 with a dot over it so that $\frac{2}{3} \sim .\dot{6}$, which is read "point six recurring." The decimal form of $\frac{2}{3}$ has only 6's but an infinite number of these. From this form we can derive those for $\frac{1}{3}, \frac{1}{6}$ and for $\frac{1}{9}$.

$$\frac{1}{3} \sim \frac{1}{2} \times \frac{2}{3} \sim .\dot{3}$$

$$\frac{1}{6} \sim \frac{1}{2} \times \frac{1}{3} \sim .1\dot{6}$$

$$\frac{1}{9} \sim \frac{1}{3} \times \frac{1}{3} \sim .\dot{1}$$

The successive steps in the calculation of the decimal form for a fraction like $\frac{2}{3}$ may be rewritten in a condensed form that suggests the steps of a division routine.

$$
\begin{array}{llllll}
2/3 & \sim & 2.0/3 & \sim & .6 & + & .2/3 \\
.2/3 & \sim & .20/3 & \sim & .06 & + & .02/3 \\
.02/3 & \sim & .020/3 & \sim & .006 & + & .002/3
\end{array}
$$

These steps may be merged in a procedure whose steps are written out in a vertical notation like that of the division of integers in the previous chapter.

$$
2/3 \quad \sim \quad 3\overline{)2} \quad \sim \quad 3\overline{)2.0}^{\,0.666} \quad \sim \quad \cdots
$$

$$
\begin{array}{r}
20 \\
20 \\
2
\end{array}
$$

The steps in the derivation of, say, $\dfrac{2}{7}$ may be set out in the same way.

$$
2/7 \quad \sim \quad 7\overline{)2} \quad \sim \quad 7\ \overline{)2.0}^{\,0.285714} \quad \sim \quad \cdots
$$

$$
\begin{array}{r}
60 \\
40 \\
50 \\
10 \\
30 \\
2
\end{array}
$$

Here we stop at the remainder 2, because we can sense that the process will repeat itself, since we started with the 2 as a dividend. But the decimal is <u>endless</u> and can be extended for as long as we wish.

$$\frac{2}{7} \sim 0.285714\ 285714\ 285714\ldots$$

The group of digits that are repeated indefinitely is called the period of the decimal. A form like that found for $\frac{2}{3}$ is also "periodic," but in this case there is only the one digit in the period; furthermore the period need not appear immediately after the point — for example, $\frac{1}{6} \sim .1\dot{6}$, which has an "irregular" part, here the single digit 1, before the period appears. We use the notation for recurrence in the case of longer periods by placing a dot over the first and the last of the recurring group of digits — for example, $\frac{2}{7} \sim .\dot{2}8571\dot{4}$.

The new way of writing fractions with denominators that are powers of 10 has been made to make sense, even when extended to other fractions. This has involved new awarenesses: that the decimal form of any fraction may be derived by a division routine; that some decimals turn out to be endless; that these have recurring digits, or periods; that such periods need not necessarily start immediately after the point.

The elementary study we have made serves to illustrate how mathematical awareness is educated: first, as awareness of language and a corresponding notation, then as an opportunity to ask questions about both the language and the notation, to see how far they can be extended. Articulating the awareness provides new mathematical openings for both students and teachers.

We now have a new type of number in the form of infinite, recurring decimals. But we also have another opportunity to extend these further, by simply dropping the proviso that there should be a recurring period in the infinite sequence. It may not occur to us to do that. But it may happen that someone who dreams about these infinite strings of digits wonders whether it would make any sense to even suggest that these sequences be totally random; to wonder whether such "monsters" already exist but are as yet unrecognized.

It may become easier to make this sort of extension after it is understood that <u>every</u> recurring decimal can be translated back into the form of a fraction, whose numerator is the period and whose denominator is the number written with a string of as many 9's as there are digits in the period. For example, the decimal $.1\overset{\bullet}{4}285\overset{\bullet}{7}$ is equivalent to the fraction $\dfrac{14287}{99999} \sim \dfrac{1}{7}$. We have not established this sequence here; but we have done so in Book VII of the series Mathematics with Numbers in Color; this contains all the necessary steps that lead to this profound awareness.

It is profound, because it will prepare students for that which became gradually clear to professional mathematicians over the last hundred years, namely that there is a class of numbers comprising all possible infinite decimals. What we have studied so far is only elementary, namely that part which does not require the fundamental changes in mathematicians demanded by so-called "set theory."

We shall develop this no further here, beyond emphasizing that it seems possible to give a new foundation to mathematical education without having to repeat the historical development of mathematics. Ignoring this development we may go directly, via language and awareness, to the way the mind works when it asks questions about that with which it has made itself familiar.

When we agree to look into decimals with an infinite number of digits, we come across the way mathematicians managed to replace an impossible task — that of seeing an infinite number of entities simultaneously — either by a device like "recurrence" or by the "unexpressed" thought that there is no need actually to know every digit, and that, instead, we can just remain aware that there is no end to the sequence. The class of digits is somehow like any class of objects which are not known, yet when one turns up it is immediately allotted to the class — as with glasses, chairs, houses, or indeed the class represented by any noun.

The transfer of this awareness — of a class associated with a concept — to the sequence of digits in a decimal may be a very useful exercise. If we forget that to every word there always corresponds a concept, and so a class, we may feel that infinite sequences of digits are insuperable obstacles. But words, when learned from the start, are conceived by babies as open — indefinite — classes. It may be advisable to take advantage of this in the mathematical education of young students.

Another possibly useful exercise is to express decimals known to have a form with few digits in equivalent forms with an infinite number of digits. For example, .5 has only the one digit, but it may be written as .500 . . . with an infinite number of zeros, or, more dramatically, with an infinite number of 9's, since $.5 \sim .$ $\frac{4.5}{9} \sim .4\overset{\bullet}{9}$. This may force awareness that we have a simple way of expressing <u>all</u> decimals as infinite sequences.

We can extend decimals in this way as objects of thought. In calculating, however, we <u>must</u> avoid anything which makes the job impossible; to keep an actual infinity in focus in this case is one of those insuperable obstacles. Mathematicians have invented "approximations" in order to surmount such obstacles. Only a small number of digits are retained, so that calculations are possible. But an extra awareness is then required to estimate the discrepancy between the result obtained in this way and the "unknowable" answer — unknowable because exact calculation cannot be done.

•

Since .49 is another way of writing .5, we can force awareness that if we only take two digits, that is .49, we can calculate the discrepancy between .5, which is equivalent to .50, and .49, namely .01; if we take three digits, .499 is .001 short of .500; .4999 is .0001 short of .5000; and so on. The word "short" is clear and it is one of the meanings of the word "approximation." At every step in the above calculations, we made the difference between .5 and the selected decimal smaller and smaller — since $.01 > .001 > .0001$.

Approximation means getting nearer; we can attach a number to this notion of "coming closer," for the numbers .01, .001, .0001 "measured" the discrepancy or difference. This allows us to force awareness of "better" approximations.

A more detailed treatment of these matters is given in Mathematics with Numbers in Color, Book VI. This shows how young children can learn to replace impossible tasks by "possible tasks," with a precise meaning given for this substitution which is called "estimation." Exact knowledge is far rarer than approximate knowledge. It is therefore useful to know, and make explicit, which sort of knowledge we are after, and why, rather than leaving children to believe that exact knowledge is the most common and the other a nuisance.

Our study of decimals has added another thing to the education of our awareness. It has brought us closer to the reality of those mental powers which are involved with the activities of everyday life where we have to judge whether we can act, while being satisfied that we are dealing with the imperfect, accepting it as if it were what is required. For example, half-apples cannot really be halves, but we know how to be satisfied that what we see is "good enough" and we move on, accepting our "half." Another illustration may be found in the way our eyes make immediate estimations to match a presented challenge: we aim at a target — a marble, a person running, an enemy plane — and an exact aim here means one that "suffices to hit."

A Useful Fraction: Percentage

Our bias towards linguistic awareness means that this section can be very short. Percentages are mainly of value in applied fields, such as economics, but we consider them here as expressing an awareness of certain fractions, those with denominator 100, for which a language was proposed and found useful.

Any fraction $\frac{a}{100}$ may be written $a\%$ and read "a percent." In effect, every percent or "percentage" is a decimal, ".0a", but a new notation is justified by the many situations that are best described in this way. For example, rates of interest paid on borrowed money are customarily given in percents. A "rate of 9.5% is not an amount of money; but it yields one — the interest — when it is applied to a specific amount. Thus, the interest from 9.5% <u>of</u> 100 units of currency is $\frac{9.5}{100} \times 100 \sim 9.5$ units of that currency — the percentage acting as an operator.

As operators, percentages can be added or subtracted on the understanding that they are applied to the <u>same</u> entity. Thus, 2% + 3%~5% in the sense that 2%×a + 3%×a ~ (2%+3%) ×a ~ 5% × a.

Percentages can also be multiplied or divided as long as they remain operators and yield percentages. Thus, 2% × 3% ~ $\frac{2}{100}$ × $\frac{3}{100}$ ~ .$\frac{.06}{100}$ ~ .06%. Sometimes a fraction of a percentage is required, so that although we use two languages we operate as with fractions and call the result a percent. For example, "half of 10%" ~ ($\frac{1}{2}$ of 10)% ~ 5%; here the half was an operator, the 10 an object, and the result is an object or an operator, according to whether the percent itself is being applied to an object.

We learn from this that it is the mental climate we associate with a problem that determines whether we merge the various ways of describing things into one entity, by looking at them in a certain way, or whether we stress their distinctness in order to serve differing purposes. We therefore need to remain alert to be sure that a shift from one climate to the other is permissible — from operators to objects, from fractions to decimals or percentages. We consider this to be the true education these opportunities offer.

The know-how of using percentages is a by-product; the challenges to one's awareness are the most important — scare of percentage I shall be left them are indicated in the following exercises.

- Read various integers as percents — for example, 1 $\sim \dfrac{100}{100} \sim 100\%$.

- Read various decimals as percents — for example, .09 $\sim$ 9%; 1.1 $\sim$ 1.10 $\sim$ 110%.

- Read various fractions with denominators that are powers of 10 as percents — for example, $\dfrac{5}{1000} \sim \dfrac{.5}{100} \sim .5\%$.

- Read various other fractions as percents — for example, $\frac{2}{3} \sim .\overset{\bullet}{6} \sim .6\overset{\bullet}{6}\overset{\bullet}{6} \sim 66.6\%$, or alternatively, $\frac{2}{3} \sim (\frac{200}{3})\%.$

Because percentages are mainly useful in applied arithmetic, they give us an opportunity to educate young students in aspects of economic life that do not generally come their way in school.

- In my country taxes are paid at different rates according to a formula; for example, 12.5% of $\frac{1}{3}$ of my income, then 20% of the next $\frac{2}{5}$, and then 45% of the rest — what percentage of my income shall I be left with?

- According to another formula I pay 50% of $\frac{1}{2}$ of my income, 100% of the next $\frac{2}{5}$, and 160% of the rest — does it still make sense to ask what with? Could such a formula be used in any country? Why not?

Infinite Sets

Traditionally, educators have been guided by historical developments, as if there was a correspondence between the mental growth of youngsters today and of mathematical thinkers of the past — as if ancient stuff must be easier stuff. In fact, it is just the other way around. Over the last hundred years, mathematicians have re-thought the whole of mathematics and have simplified a great deal of it. It is therefore possible to recast curricula to the benefit of learners at all levels. They will then be able to join the ranks of mathematicians, since they will know the meaning of mathematization in many areas. We shall consider another such area, taking as an example a "modern" topic which has profound meaning but which is simple enough to be understood and enjoyed by young students.

We already know that the sequence (I) can be split into sets of integers, the odd ones (O) and the even ones (E). It is obvious that there are "as many" numbers in (E) as in (I), since for every n of (I) there is a $2n$ of (E), and conversely. Another awareness lets us also know that there are "as many" odds as evens, since both sets were obtained from (I) by retaining only alternate numbers.

If start with (0) and double each number in it, we obtain a set (E_1) of integers that are all even, but not including all evens — for example, 4 is missing.

$$(O) \quad 1 \quad 3 \quad 5 \quad 7 \quad 9 \quad 11 \quad 13 \quad \ldots$$
$$(E_1) \quad 2 \quad 6 \quad 10 \quad 14 \quad 18 \quad 22 \quad 26 \quad \ldots$$

If we double the numbers of (E_1), we get another set (E_2) of even numbers. Here too it is clear that there are even numbers not contained in either (E_1) or (E_2) — for example, 8 and 16. Doubling again will yield another set (E_3), and this process may clearly be repeated.

$$(E_2) \quad 4 \quad 12 \quad 20 \quad 28 \quad 36 \quad 44 \quad 52 \quad \ldots$$
$$(E_3) \quad 8 \quad 24 \quad 40 \quad 56 \quad 72 \quad 88 \quad 104 \quad \ldots$$

$$\ldots\ldots$$

Various observations arise from an examination of these sets of even numbers.

- The subscript attached to each letter E naming the set, indicates the number of times (O) has been doubled to obtain the set — once for (E_1), twice for (E_2), and so on.

- The difference between two successive numbers in (E_1) is 2, in (E_2) is 4, in (E_3) is 8, and so on.

- When we write the first few numbers in each sequence, we know how to continue; but we do <u>not</u>

do so, because if we did we would get lost in infinity.

- Once we know that doubling each number in a previous sequence gives us the next set of integers, we see that there is no "end" to this process.

Thus, we become aware that the set of integers (E_n) is made up of an infinite number of infinite sets, each of them having "as many" integers as there are in (O) or (E). This <u>new</u> awareness may be an eye-opening — mind-boggling — experience, but it is also an example of a mathematical fact, requiring no knowledge other than what has already been met in the previous chapter.

This first contact with what is called a "countable infinity" may appear to some teachers as an isolated mathematical fact with no immediate significance. But for those concerned with the education of awareness, it will provide a wonderful opportunity to open up vistas to students, whose minds are so open to experience, and yet who in the past have been so often simply regimented and kept on an uninspiring mathematical diet.

The answer to the question "how many?" is usually found by counting. We cannot actually count these infinite sets; but we can answer the question satisfactorily by saying "infinity" or, sometimes as we already have done, "an infinite number."

Actual counting requires that a set be "finite," the meaning of which is that the set can actually be counted. This is a circular definition — there are lots of these in mathematics. But it is not circular in terms of awareness; for "actually counting" means that the action can be in fact performed, even were it to take a long time, whereas in mathematics, "actual counting" is replaced by a "virtual" operation — one which says "yes" to the question "can it be done?"

Actual operations have to become virtual, because of the presence of infinity in every mathematical thought. In defining numerals as adjectives, cardinals as enumerators of sets, numbers as finite or infinite equivalence classes, we replace the actual by the virtual. We act as mathematicians, who are also involved in the demands of life and aware of practical limitations. Like them, students can come to know whether an additional component — infinity — is part of their thinking or whether they are only concerned with operations and operations on operations. Our students are entitled to be given the chance to prove to themselves that they can behave like mathematicians, even though their main concern at the time is not to produce theorems.

Summary

This chapter has been a development from the previous one, though both aim to force awareness that mathematization is a special way of being that is open to all.

Any alternative curriculum may still be treated as a body of knowledge that has to be passed on. But we insist on the <u>little</u> that needs to be given by teachers, and the <u>lot</u> that students can invent by themselves, once their awareness has been forced and enough practice given for them to feel in control.

With awareness comes certainty; for students now recognize that there are real criteria for deciding things, not just a number of seemingly arbitrary activities. They will know what these criteria are and can be guided by them as they progress in their study of mathematics. It is in fact reasonable — and also important — to make students <u>independent</u> of their teachers and textbooks, <u>autonomous</u> in their work so that they sense their own initiative and see it at work at every stage, and <u>responsible</u> for everything they do — this being a separate awareness developing out of the other two.

Mathematicians achieve these aspects of their activity after much work on themselves and in collaboration with their peers in particular fields.

We can make students mathematize again and again, rather than teach them mathematics. Having seen how this can be done in a number of areas, we can say — with precise meaning — that we make mathematicians out of our students. In doing this, we remain in contact with those parts of classical mathematics that are still retained in the traditional curriculum; but we inject these with many new insights, feelings and awarenesses.

It is in this spirit that we invite teachers of elementary mathematics to examine carefully our proposals.*

─────────────────

* See Appendix for descriptions of **Student Sheets** designed by author to introduce many of the mathematical awarenesses of this chapter to students.

12 Educating the Whole Brain

Introduction

We have concentrated on language and algebra in the last two chapters and, so far, we have deliberately avoided any reference to the dynamics of imagery or to what is called space. Here, in this chapter, we will be concerned with these mental dimensions — with perception and geometry.

School education has always been mainly verbal. When geometry was a subject in the curriculum, many students could not take to it easily and developed the feeling that it was not for them. When — twenty years or so ago — scientists began to refer to the different functions of the two hemispheres of the brain and the dominance, in most people, of one, then it seemed that here was an explanation for the deficiency of geometrical intuition in so many, otherwise good, students.

An over-emphasis on verbal and logical powers of the mind at the expense of others would indeed mean that any subject that invoked neglected powers would be found difficult and demanding. There is a certain creative aspect of geometry that requires all of oneself. To be geometers, students have to be inventive, in contrast to some extent with algebra, which — once mastered — can be made mechanical, and so only demands part of the self.

Much of our school geometry has been inherited from the ancient Greeks. They turned to geometry as a way of handling the challenge of "irrational numbers," of which $\sqrt{2}$ was a disturbing forerunner. But in doing so, they gave geometry a formal, syllogistic unfolding from definitions and axioms to theorems. This approach dominated mathematical thinking for 25 centuries. Euclid's famous treatise remained a model for the rigorous treatment of an exact science during all that time. Moreover, it became accepted for some time that mathematicians presented geometric <u>results</u> in a logical way that was quite different from the more "intuitive," and undescribed, <u>methods</u> which were used to establish the results.

Later, mathematicians continued to try to find systematic approaches to geometry. In the 17th century, Descartes developed "analytic geometry." This was a way of treating geometrical problems by the manipulation of algebraic equations. His ideas were immediately accepted because there was such an obvious difference between purely geometric solutions, which seemed to require something akin to genius, and this new approach which could be made mechanical.

It is only relatively recently that these classical presentations of geometry have been felt to be too narrow. New proposals have come from some emancipated minds who wanted to save what was educationally valuable in geometry by integrating as much of it as possible with other developments. One of these developments is topology, which has been a thriving branch of mathematics in this century.

In general, mathematicians tend to work in some preferred branch of mathematics. Those who are more at ease with manipulating operations are called "algebraists;" those who are more at ease with manipulating sets are called "topologists." More recently, some mathematicians have been able to integrate naturally both approaches and these are called "analysts." There may be a temperamental basis for such distinctions; in other words, it may be that it is the soma and the brain that are the deciding factors in anyone becoming a professional mathematician.

When we consider such matters, we clearly go beyond the classical view that, because of its peculiar nature, mathematics has an absolute existence per se, independent of human minds. We may now see it as made up of the many products of different minds. People have different minds because of their make-up, and because of what they do to themselves from conception onwards. They are affected by the particular way they acquire their experience — of the world and of themselves — and find unwittingly that this involves a great deal more than the intellectual components of mathematics. Some aspects that are involved are: <u>temperament</u>, which describes physical make-up, including the brain which is the unique creation <u>in utero</u> of every human being; <u>affectivity</u>, which includes taste, pleasure, and motivation to pursue one's engagements, in certain directions, in certain ways; <u>ambition</u>, which indicates the projection of oneself in society, to achieve whatever seems possible with one's gifts; and many more than these, all interacting in the actual unfolding of one's life.

Mathematicians mathematize; this is an overall activity involving as many aspects of one's self as are present. These need to be discovered and studied in order to yield the knowledge that can help produce the proper mathematical education of all schoolchildren. For all mathematicians have been children, and their own spontaneous education, little known to their parents or teachers, was in their own hands. On their own, they selected awarenesses which stayed with them even when forgotten. These awarenesses guided them in their choice of how to consume time in their life, and — at some stage — in their choice of what kind of mathematicians they wanted to be.

* * *

Our experience is far from random, although many unregulated impacts find their place in it. Our experience is structured, both by refusing room to some of what reaches us, and by the fact that some experiences must precede others in time — as standing has to come before walking, and walking before running. This necessary, hierarchical sequencing is found, for instance, in the presence or absence of interest expressed by all children when they get into, or abstain from getting into, certain activities, at certain ages.

A more detailed development of this theme may be found in the books, The Universe of Babies and Know Your Children as They Are. But enough has been said here to indicate that <u>perception</u> is an "absolute" for all of us when we are very young children. This is soon assisted by <u>action</u>, so that our "need to know" the environment around us can be pursued and satisfied. Action now becomes the next absolute; it is now assisted by perception, allowing us to know ourselves-in-the-world. This ends when action is extended to being "virtual" and so a basis for thoughts and ideas, which may form an absolute of their own — the absolute of <u>intellect</u>, strongly developed in mathematicians and philosophers.

But this commitment to intellect only takes place after the years each of us spends in knowing our inner lives, woven from the subtle substance of feelings and sentiments, based on emotions. The absolute of adolescence integrates all we did earlier, in order to extract the ingredients of the mental powers that enable human beings to transcend the constraints of any environment, to reach the personal, inner universe characteristic of every individual.

To understand any of the many forms the "need to know" takes in the lives of the human beings around us, we need a proper understanding of the temporal structuration of experience. Its importance is underlined by the way we try to reach, in this chapter, a proper education of the whole brain. We do this by providing experiences which are natural within each successive absolute, and ending with ways of integrating these experiences into one specific form, that displayed by mathematicians who are analysts.

What this means is that algebraists can as easily be topologists— and conversely — because their education has forced awareness of what mental powers have to be used for specific purposes and what it is that can make them as strong as possible.

In the following sections, we set aside global intuition of human experience and knowing, to develop the specific geometric exercises which become, in time, the know-hows of working mathematicians.

Perception and Action

When the self dwells in a particular absolute, the need to know manifests itself with passion. There is a zest to learn, found in all the experiments the self gives itself spontaneously, often for hours on end.

Sometimes, certain activities can be chosen which, while nourishing that passion, may also yield knowledge and know-how that is also valued by society. Such activities were proposed, for instance, by Maria Montessori at the turn of the century. Her structured teaching material fascinated many young children and continues to do so today.

The way in which the powers of perception and action can come together may also be seen in the use of Cuisenaire rods. These were invented by Georges Cuisenaire to help so-called "slow learners" — which they did remarkably well. Their true value was revealed when they were seen as being able to offer insight into spatial relations underlying arithmetic. A set of Cuisenaire rods was "a spatial model for the algebra of arithmetic." More of the brain was involved in this approach, which until then had been merely verbal and founded — mistakenly — on cardinal counting.

The rods could be obviously increased in length by placing them end to end. Addition became visible without counting, as was subtraction by covering part of a rod by a smaller one — "taking away" what was covered, and at the same time making it plain which was the part left over. Classes of partitions became immediately understandable through exploration of equivalent lengths made with different rods. Images rather than words were able to accompany notions like the commutativity of addition, the distributive law, multiplication, powers, and so on.

The details of how this is done are available in a number of publications which have made available to young children a new curriculum in which "algebra precedes arithmetic." The obviousness of this order of mental structuration had escaped all who paid tribute to tradition — everyone, in fact. But the rods did not make the expected spectacular transformation of the teaching and learning of mathematics in elementary schools throughout the world. This is because they were seen as nothing more than "manipulative materials," capable of yielding everything that "society" demands should be passed on to the next generation. The underlying notion that here was an education of the whole brain did not attract the attention it deserved.

* * *

The whole brain is more obviously involved in the study of geometry. So, here in this section, we shall describe work with Geoboards in order to illustrate how we can educate perception and action to force awareness of mathematization in young children.

Geoboards are materials that are lightly structured by lines, lightly incised in a plywood base, and by some regular arrangement of nails. The nails are used to stretch thin colored elastic bands; the lines provide a background for the shapes made by the bands. The most suitable nails are so-called escutcheon pins — different boards are often referred to by the number of pins they use. The elastic bands may be of any appropriate size; colored ones are more suitable since they are more easily distinguished and described.

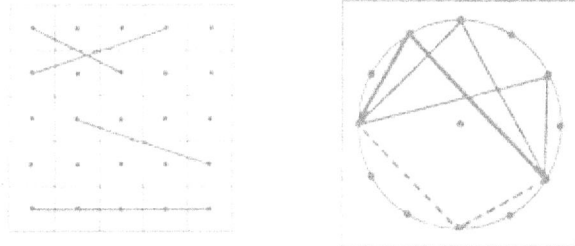

Two types of Geoboard were proposed when they were first introduced in 1953.

- 9-pin, 16-pin, and 25-pin square Geoboards have background lines incised to from a square "grid" with nails at the center of the equal squares of the grid so that they form a 3×3, 4×4 and 5×5 square "lattice," respectively.

- 6-pin, 8-pin, 10-pin, and 12-pin circle Geoboards all have circles of the same size incised on the square wooden base; the indicated number of nails are equally spaced on the circumference of the circle.

Geoboards involve two activities; one is physical, that of stretching rubber bands around nails; and one is mental, that of relating to the result of such action, and sometimes naming it, if by chance it already has a name. How much can be done with so little cannot be known a priori. But years of playing with Geoboards have led to the finding that there is indeed a lot that can be made visible and self-evident. Some activities yield the straightforward propositions found in geometry textbooks, so that Geoboards have become popular with mathematics teachers.

Some exposition of what can be done with Geoboards have been given in various publications, listed in the appendix. What we want to do here is to consider more closely what is involved when young children play with Geoboards, at home or at school.

* * *

The basic element involved is the segment of a straight line. There are then various opportunities available. The presence of a certain number of nails, located at certain positions on the board is such an opportunity. The need to know, sometimes called curiosity, is another. These create other opportunities:

- structuring plane space by using one or more rubber bands which can be placed around some nails to generate the unknown or the unexpected;

- rotating the board, held in one's hands, or placed on a table;

- adding or removing a band, or stripping the board of bands to start again;

- arranging bands in aesthetically satisfying ways;

- finding and talking about topological, as well as metric, relations, with statements about particular facts, whether obvious or otherwise, and about their connections with other facts — those also seen in the situation and those recalled from memory;

- encountering difficult challenges that no one may have yet solved — some may turn out to be insoluble in this context, opening up awareness that other lattices, or other material, may be required;

- working on the verbal statements offered by students, to neighbors or to the class, in order to chastise the language used and create habits of careful, unambiguous expression;

- creating a store of images — static when the work is done and becoming dynamic, both at the time and when the work is returned to in the mind;

- developing a sense of constant growth in knowledge of what is being done, can be done, remains to be done, and reclassifying such knowledge;

- making definitions that are clearly grasped and verbalized;

- extending basic, recurring propositions to become theorems which may be able to be proved valid for extended lattices, with varying degrees of certainty.

There are other observations to be made from playing with Geoboards, but these will suffice here. What matters is that we can keep learners very close to everything that is being done, involving them affectively, perceptively, actively, verbally, in fact totally, in their contemplation of a challenge. Rather than trying to derive well-known properties in order to memorize them, we keep students on the edge of the unknown, near their own unfolding of propositions which stem from themselves. Such statements will be functionally interrelated and rooted in simpler structures. The students will know exactly when these statements belong to topology and when algebra is present.

They will know when the local suggestion on the Geoboard can be given universal significance, so that the words used to speak of it are sensed as global.

All students will be able to speak spontaneously of these matters as if they were as real to them as other objects, and in a manner which makes sense to them. In other words, we can educate the geometer in every student. This demands the use of eyes and hands, of will, of feelings and words, and of easily recalled experience. So when we hear the students making statements that are recognizably geometrical, we can also know that we are educating the whole brain.

Various specific geometrical awarenesses will be available to the student:

- segments on the same line can be added and subtracted;

- lines can intersect or be parallel; intersecting lines make angles — called right angles when the lines are perpendicular;

- lines can bisect angles equally and can bisect segments equally;

- segments can form sides of convex, concave, or overlapping, polygons — adjacent sides of polygons intersect at vertices, lines joining vertices not on the same side are called diagonals;

- polygons have as many vertices, or angles, as sides and are named after this number, using Roman prefixes in the case of tri-angle and quadri-lateral, followed by Greek ones for penta-gon, hexa-gon, and so on;

- some segments are not so easily measured in terms of the side of the unit square — for example, the diagonal of unit square itself.

The experience of the student is not developed here as it is in a formal logical presentation of geometry. It is empirical, stemming from various roots rather than one single system of definitions and axioms. That is why it is called "Geoboard geometry." Students start from something perceptible and label it with the help of the teacher; further developments continue to be based on what can be perceived. For example, the classification of quadrilaterals could start from the most general — about which nothing can be said other than that it has four sides — and then be progressively structured through parallelism of opposite sides, equality of pairs of sides, and equality of diagonals, with the resulting categories being given their traditional names. Such a classification relates shapes that might otherwise be presented arbitrarily, and does so now in perceptibly clear ways.

* * *

Geometry need not be a strictly deductive system but rather a vast net of experience which one day may become organized that way. Such an end-product is less valuable than the structuring of the student's mental vision in order to engage in dialogues, stating "geometric facts" as propositions known to be true for a number of reasons, that are not necessarily deductions from a single source. Geoboard geometry cultivates creativity, autonomy, ease of expression, accumulation of experience, and various know-hows valued by mathematicians. In a word, it teaches mathematization.

Mathematization is not only accompanied by the joy of discovery but also by functional retention. Indeed, retention, rather than memorization, is the basis of this experience. There will be no need for memorization, except in so far as ogdens have to be paid to remember labels required to standardize descriptions. Diagrams will be generated by students; they are easily retained through the experience of having constructed them step by step. Students will do and undo, showing and verbalizing as they go along. Geometry will be action — actual at first but soon virtual, as it is for mathematicians.

Propositions sum up experience. They can only appear at the beginning when a subject is recast in order to yield further results. Propositions are like the statements of everyday life;

they lead to conversations between people. They are concrete, familiar to the speakers and accessible to the hearers. They relate to one's mental experience, to one's images, like everything else that is mastered by being lived. They seem to be retained in the same way, being recalled spontaneously and smoothly in the way images are.

That is why we say that we educate the whole brain.

Complexity exists — it has its place. It may be distinguished from complication, which we experience when randomness intrudes into our thoughts. Complexity is much more natural in life than simplicity; it gains its proper place in the present context. For playing with Geoboards is not just a matter of making obvious those geometric propositions that are found in textbooks. It can be one of those rare situations where "a little produces a lot" and where "nothings" are so clearly important and powerful. Mathematicians do both these things, and now we can make most children own some of them, intimately and naturally, a true gift for those who want to feel mentally more powerful.

Animating Space

Students acting on rubber bands or colored rods see their actions as discontinuous. They structure their space through their own actions and perceive it as having attributes connected with their way of approaching it. The precise content of their minds is hard to know.

Another approach might be to construct ready-made images to be offered directly, so that students still use their perception, but their action is now all virtual. A way of generating such images was proposed by Jean-Louis Nicolet in 1940, when he made films of animated diagrams that could be projected and viewed on a screen.

Working alone in Switzerland, and with meager resources, Nicolet produced short, black and white, silent films, lasting from one to three minutes, whose striking beauty accounts for much of their appeal. Seeing the films has always been appreciated as an aesthetic experience by all viewers, whatever the mathematical background they bring to them.

As a secondary-school teacher in Lausanne, Nicolet was concerned that so many of his students lacked the preparation to be able to love geometry as he did. He worked hard at animating some of the diagrams from the geometry course he was supposed to teach, and found the result effective. Not only did his students understand the theorems, and could learn things that had been too difficult in the past, but they asked for more, asserting that this way of meeting geometry made sense to them.

Nicolet died in 1966, having made a number of films at his own expense, and having been almost totally ignored by mathematics teachers, in his own country and elsewhere. The exception was a handful of keen reformers who appreciated his work but did not manage to make it more well-known, finding — like he had — that very few were sensitive to his contributions.

A London college teacher, Trevor Fletcher, had independently produced much longer and more sophisticated mathematical films, animating diagrams to generate a number of theorems and their proofs using only visual means. Despite having a wider audience than Nicolet, he, too, remained an isolated film producer for some time.

The author joined this handful of filmmakers by producing three short films in 1958. These were followed by other films in the 1960s, under the title Folklore of Mathematics, to counter the then prevailing reforms instigated by some mathematicians in the name of "modern" — or "new" — mathematics.

These reforms lacked sensitivity to mathematization. In most countries, people adopted new content but continued to teach in the ways that had already failed, ways that were exclusively verbal and symbolic. One conclusion drawn from the inevitable failure of these "reforms" has been a resigned, even if sad, acceptance that mathematics is not for everyone, but only for the "gifted." In particular, this has been the fate of geometry in the Euclidean form in which it is usually cast in schools.

Readers will know that this is not the conclusion of The Science of Education. Certainly, straightforward formal presentation of chapters of geometry — or any other branch of mathematics for that matter — does not yield results for the majority of the school population. But the development of a sense of mathematization does. When the powers of the mind related to imagery and imagination are better known and used, then students not only involve themselves in geometrical challenges, but enjoy them: geometry makes sense.

* * *

Many experimental lessons using films have been conducted with classes of adolescents, often as many as a hundred or so in a large room. On these occasions, it has been clear that almost all the students were involved and could discuss with each other what occupied their mind after a viewing of the film. When it was impossible to get unanimity on what had been projected onto the screen, a second viewing — or a third — enabled doubts to be settled. During the discussion, students would be invited to recast their statements so as to eliminate any misleading ambiguities; sometimes technical terms would be suggested and these would be willingly accepted and used.

Such experiments establish various findings.

- Animation involves a deliberate use of the "continuum," easily followed by the eyes which also move continuously leaving behind an infinite number of interlinked images that yield a "visual statement."

- The smooth and pleasing movement of diagrams on the screen generates an "affective charge," which can last for some time, and which holds the images together, so that speakers can speak of them as they would of a landscape.

- These images impose their own "reality," so that viewers believe that any particular statement about them will be "true" and "not true," sometimes immediately and with total conviction. This is the basis for the certainty needed to state that things are "proved" beyond any shadow of doubt. Later that feeling will be transferred to verbal and symbolic proofs, capable of triggering the "reality" behind them.

- What is contemplated can generate more than what was seen; in other words, it has a future and can perhaps lead to further geometric creativity.

By creating a new format for lessons, in which "dialogues" are the rule rather than a passive acceptance of what someone says is important, we stay close to the realities of the mind as well as to mathematization. It is clear that such an education is going to command students' assent, mobilize their energies and perhaps generate enthusiasm. To secure this, we can take advantage of the aesthetic qualities of well-made animated films, which are nowadays much easier to make using computer animation techniques.

Such films have been created primarily to force a number of awarenesses, not least that "I too can think of space creatively."

There is a duality in the language of animated film. On the one hand, it tells "stories," recounts events which can be witnessed, recalled, and talked about. On the other hand, it allows our minds, which have time as one of their attributes, to be with the film and to identify the dynamics of the kind of images found in the dialogues that geometers have with themselves. In doing this, our minds mathematize reality.

Circles in the Plane

Some of Nicolet's themes have been treated in a new series of geometry films. These are computer-animated, color films, lasting about six minutes each; further details are given in the appendix. We will give a detailed analysis of one of these films, Families of Circles in the Plane, as a particular example of how our approach can recast traditional topics into a more integrated and accessible whole.

Anyone who can trace a circle on a piece of paper with a pair of compasses knows that a large number of them can be drawn. One only has to choose a point of the plane on which to place the sharp point of the compasses and then choose an opening for the arms in order to trace a circumference. The two variables — called the "center" and the "radius" — chosen in this way are felt by the active and perceiving self as directly knowable, as real as one's hands or eyes.

The first transformation of awareness may consist in seeing that <u>all</u> circles of the plane can be generated, individually and as a set, by acting on these variables. Thus the film starts by showing one circle — in red, an arbitrarily chosen color — moving in the plane, that is, changing its center and its radius all the time. If we stop projecting the film at this point we can require students to become aware that no circle in the plane can escape being generated in this way, and that one needs to think simultaneously of the centers and the radii as variables in order to take in the whole family.

After such a start in structuring a dialogue, we can feel free to consider individual circles which are then singled out as members of possible sub-families though these have not yet been defined.

*   *   *

A second awareness may be forced when the film now lets a white dot appear, so that there can be a suggestion of a relationship between this fixed point of the plane and the family of circles which are held at the back of our minds. One may ask what happens to the family when the relationship chosen is that the point lies on the circumference of the circles of the family.

There will then be a partition of the whole family into circles whose circumferences pass through the point, and those which do not. the film focuses on the first subfamily, by coloring in green — also an arbitrarily chosen color — all the circles whose circumference passes through the white point. Such circles may still have centers anywhere in the plane, but then, in order to pass through the white point, their radii can no longer be arbitrary — they are "bound." That is why the red coloring was abandoned and the green introduced. An examination of this situation shows that the new family is defined by that which constrains it. We can put into circulation the phrase "degrees of freedom" by saying that the circles have lost one of these degrees.

*   *   *

There was no need to say too much about the red family, its main point lay in its existence. Not much more need be said about the green family. To articulate a clearer and richer dialogue, we cut into the green family by imposing further restrictions on the circles — for example, making all the radii the same size. This is done briefly in the film, and then these circles become blue.

We can, of course, stop the projection at this point and dialogue with the situation now generated. For instance, we can make the following observations.

- By choosing the common radius of the blue circles to be any size from zero to infinity, we generate a family of subfamilies which will cover every one of the circles that were previously colored green, those whose circumference passed through the white point. By this device, the green family is structured into a family of circles.

- The subfamily of blue circles covers a "disc," centered on the white point and with a radius that is twice that selected to define the subfamily. By varying the radius, such discs, which are seen only in our minds, can be made to cover the plane, to generate a new subfamily — a family of "concentric" circles, centered on the white point.

The film shows another blue subfamily, made out of all circles which remain "tangent" to a fixed line — which is not shown — passing through the white point. There are two awarenesses associated with this section of the film:

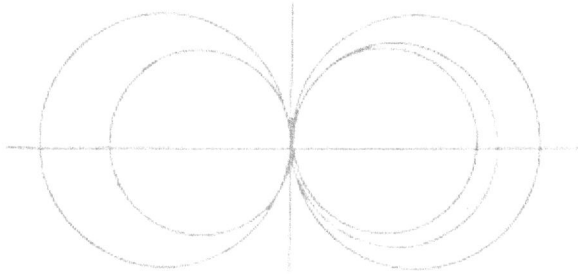

- The centers of all these circles lie on a line that is "perpendicular" to the invisible "tangent" line through the white point;

- Any line passing through the white point may be chosen as the invisible tangent line, so determining other subfamilies.

When a second white dot appears somewhere else on the screen, we have an opportunity to examine — visually at first, with the film, and then verbally — the subfamily of circles in the plane which are made to have two points in common. The dynamics of the animation permit the following observations.

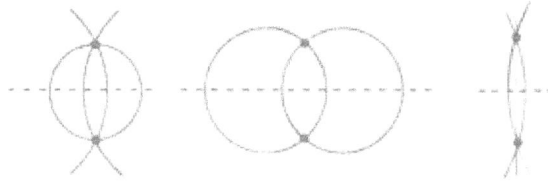

- One special circle of the subfamily is the one having the segment joining the two white points as its diameter.

- Two equal circles, one on either side of the line through the white dots, structure this family whose centers are all on the "perpendicular bisector" of the segment joining the two white points.

- Another special case of this family is then the line going through the two white points. This has two "lips," according to whether we see it as one or the other of the two "symmetric" circles of the equal pairs considered above.

By this stage, we have been able to force awareness of pairs of equal symmetric circles; of straight lines being considered as "circles of infinite radius;" of "points at infinity," one or two of them according to need; of one circle seen as two coincident circles; of axes of symmetry; and of "singular cases" among the circles of a subfamily. Moreover, the film now generates multiple visions of the family through a "ballet dance" of two circles, on each side of the line through the white points. Such moments are especially powerful, because it is our vision that is being educated; a vision which we can use on other occasions, rather than the facts it can lead us to in the form of the particular geometrical content of the images seen.

* * *

A theme filmed by Nicolet — "three points determine one circle" — is now taken up, as an incidental moment, realized by adding a third white dot and now coloring white those previously blue circles which also pass through this point. The screen is reduced to one circle — in white — passing through the three white points. The indefinite article "a" has become the numeral "one." From now on, "white" means something fixed, static, a given we want to keep as it is.

The significant progress lies in knowing that to single out one circle in the blue family requires another — a third — point. There are an infinite number of such possible points on each circle; any of these will determine the "same" circle. A link is made between the entity "a circle of the plane" and "all its points," three of which are sufficient to define it. Not much is made of this awareness in the rest of the film, but it may encourage teachers to throw their students into this movement, holding a single object and an infinite number of others within the same thought, and so providing, if need be, a shift to some other awarenesses.

The next section of the film provides a new point of view. Though circles remain the main moving elements of the screen, their centers are now also shown, so that two languages become available to the viewers. The white circle and one of its diameters — also in white — will be the "frame" for the dialogue that follows.

Two blue circles touch the white circle at one of the points at an end of the white diameter; their centers lying on the line containing the diameter. These two circles generate a subfamily of blue circles. We can say <u>new</u> things about these circles by describing them in terms of their centers. For example, we can choose pairs of centers so that as one blue circle gets bigger, the other gets smaller. We can also choose that when the latter reduces to the point at the end of the diameter — a circle of zero radius — the corresponding larger circle is then a straight line, namely the tangent to the white circle at that point. In this case, all the other pairs of blue circles will lie between these two extremes.

But the pairing may be more specific, we can, for instance, see the situation in terms of the distance from the centers to the point of contact of the circles, and "sense" a relationship between these distances. Calling the distances x and y, such a relationship might be written, $y \sim \frac{1}{x}$ or $x \sim \frac{1}{y}$ or $xy \sim 1$. The film also shows the relationship $xy \sim -1$.

These awarenesses may open up a very different dialogue — about functions, or mappings. It is easy to let students "dream up" relationships linking the distances between the centers and the white point of contact. Even if nothing is done at this stage with these, an opening has been secured. The vagueness of the relationship in the film is equivalent to its being as rich as it can be made to be.

* * *

The idea of having one blue circle tangent to a given white one lets us choose new subfamilies. A red circle becomes green when it touches the white circle; its center will still be kept moving and its radius changing, though these variables are now not independent. To stay green, the point of contact must also be kept moving on the white circle; the green circles can be touching "externally" or "internally." A subfamily of the green circles becomes blue as soon as the radius is kept constant. This hints at another theme treated by Nicolet — equal circles touching a given circle to generate a "ring."

Every subfamily generated by varying the radius of these blue circles will "behave" like that shown on the film. There are then two important extreme cases, not shown in the film, but easily developed in the mind. One will be the family of points lying on the circumference of the white circle; and another will be that of the straight lines touching that circumference. This "duality" is a profound awareness for mathematicians, now available to all viewers.

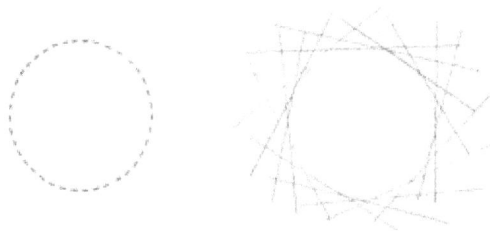

The final two sections of the film are given over to forcing some specific awarenesses arising from the simultaneous consideration of three families of circles each tangent to the white circle. Computer animation allows things that otherwise could never be done in classrooms. Each family of a green triplet can be handled separately as sub-subfamilies of circles on the plane, forcing awareness of the wealth of the original family and an awareness that our mind can be present simultaneously at various places, handling very different questions.

In this section, the white circle is not a barrier and green or blue circles can cross its boundary to unify subfamilies in the "interior" with those in the "exterior" of the white circle.

We can focus on the radii of the green circles and note that the triplets then include circles which are tangent to each other as well as to the white circle. They also include circles of zero radius, in which case only two green circles are actually visible on the screen, the third being held in the mind as a circle of zero radius. The triplet even reduces to one green circle when the circles catch up with each other and are then animated at the same speed.

As each green circle is given a fixed radius it becomes blue. The finale to this "symphony" now uses the power of computer animation to combine a number of pairs of blue circles touching the white circle at three given points. These circles have only "one degree of freedom," but that is enough to provide a "fireworks display" of moving circumferences on the screen. As the animation speeds up the generation of the blue families at the three fixed points, we gain an impression of the available wealth in this situation, and its complexity. But this does not overwhelm us; rather, we are delighted and ready for more.

* * *

In about six minutes, the film offers an experience of what geometry may be about. The stress is on geometric experience; the yield is remarkable. Students may not suspect that their elders could have missed all this wealth and beauty when they studied geometry. Having seen the film, their minds are full of images — dynamic images always open to a promising future. Geometry gains its freedom from the straightjacket of deductive systems, which reduce truth to premise, whose origins seem arbitrary and unfathomable. The most important of such systems seem to impoverish their content. Here, however, we increase the number of possible propositions that can be asserted by increasing the number of structures seen as compatible with each other and belonging to a very wide universe — the family of circles of the plane.

The Continuum

The theme of the film described in the previous section is taken up in a series of six further films, which may each be considered as a study of some family of circles in the plane. The films are not sequentially linked and may be viewed in any order to yield different things. We shall not analyze these in any further detail here. We wish rather to draw attention to the way that the many awarenesses forced by these films arise from situations involving entities like circles, rather than the points, lines and planes that are the "undefined notions" traditionally taken to be the basic elements of elementary geometry.

The films may be made to yield various traditional theorems about circles — these are almost casual by-products. But, at the same time, the films also permit entry into some relatively more sophisticated geometry. Much of this was created long after the elementary circle theorems that have always dominated geometry courses for young adolescents. To recast the curriculum in terms of awareness means that we may have to recast the importance given to some traditional topics, and to change the order of presentation suggested by the history of mathematics.

Historians of mathematics know who proposed what, when and where, they also know that new developments depend on some singular people having certain insights and suggesting to colleagues that things could be different if new attitudes were adopted. Such "turning points" are often — rightly — presented as the most valuable moments in the history of mathematics.

Implicit in this approach is the notion that the history of mathematics is a history of the special awarenesses gained by individual mathematicians as they have reflected on their own thinking. But, as we have seen, this sort of reflection can be encouraged in all students as a matter of course. The Science of Education allows us to recast the curriculum in terms of the education of the next generation, rather than preserving the products of previous generations as an unchanging body of information that needs to be passed on, intact and inviolate.

History can be transcended once we see that unnecessary, elaborated stuff can be replaced by a much more condensed vision in which all that stuff reappears in the form of special cases seen under certain lightings. We have shown how the medium of animation has been able to make such condensations.

This distillation of inherited knowledge is economic; it means that we need no longer reflect a chronological point of view. Because only awareness is educable, it seems common sense to give students the full use of their awareness — and hence of their brain — as long as it can be proved to be as fruitful as any other way of integrating a collective experience valued by the collectivity.

* * *

Films unfold images in time. Though time is <u>not</u> a mathematical notion, it is axial to all mathematical thinking and we give it a right of place when we aim at making creative mathematicians of our students. This is why the medium of film is to be preferred to any other. It is certainly preferable to the spoken or printed word. We have indicated how we educate the whole brain by letting images and their dynamics be the forerunners of words. Verbalization about one's experience after seeing a film is valuable. But it can only translate in a limited way the ineffable experience of infinite classes of simultaneous impressions gathered from the animated entities on the screen. Infinity is now actual; it is only latent when language is used.

It is difficult to reveal the wealth of geometric experience without this explicit presence of infinities. And these do not have to be treated only in the sophisticated manner of advanced courses. For instance, although the "continuum" is a transcendental notion for the classical mathematician, it is a primitive experience for every one who moves about and notices variation in the landscape. So we feel free to offer it straightaway to students, of any age, postponing for special occasions the effort of imposing onto this experience the artificial cloak of an axiomatic system.

It is only an accident of history that geometric experience has only been considered valuable when presented with syllogistic proof. Billions of people have been exposed to this — for millennia. But only a very small fraction has been able to claim to be geometers. We have seen what a vast experience can be gained by starting with the continuum and offering ways of making statements about it. So now we can let everyone choose — knowingly — whether this is exciting enough to devote oneself to the pursuit.

It is now possible for the first aim of geometric education to be the pursuit of wealth. Since we can associate other mental functions with the intuitive learning proposed, it will not be a handicap to be also offered, at some later stage, a new technique leading to axiomitization. It is far easier to superimpose on top of the wealth the formalization that is valued by some people — who may be the arbiters of society — than it is to educate creativity in those who began with formal systems.

In fact, there does not seem today to be a choice in these matters. The dismal failure of formalization in the early stages of mathematical education is a salutary warning. The proposal of The Science of Education is that we "play ball" with the creative minds of the young, and to mobilize their whole mind — and whole brain — to meet geometry face to face, making it an integral part of their mathematical behavior.

The Star

The scenario of a film called The Star is worth considering briefly here because of its simplicity and its fecundity.

The "star" is defined as a point in the plane, together with all the half-lines of the plane which have that point at their end.

Any one of the half-lines can represent the whole set if we allow rotation round the given point as a process that "sweeps" the set. There are two senses of rotation — clockwise and counter- or anti-, clockwise. If we do not stop after a complete revolution, we "cover" the set, partially or wholly, and this as many times as we wish, clockwise or counterclockwise. An example of such covering can be seen on a watch or clock with hands, for a hand can be imagined as thinned down and extended to become a half-line.

A watch may have three hands, covering the same "surface" at different speeds. In advanced mathematics this surface is called a "Riemann surface" after the nineteenth-century German mathematician, Bernhardt Riemann, who was the first to be aware that such a surface could be conceived and used to make some difficult challenges more intuitive. The surface itself is so easy to picture that it does not require anything that children of school age do not have at their disposal.

We shall not be considering the complex problems solved by Riemann. But we shall make some elementary applications — for example, in trigonometry — and this is why we take up this aspect of the star. The value of such study is to open the minds of students to possibilities, illustrating the phrase "there's more to this than meets the eye."

Of course, there are many more familiar elementary awarenesses that arise from this line of attack. We shall only list a few of the possibilities very briefly here.

- A point on a half-line which is given a complete revolution generates the circumference of a circle. So, by varying the position of the point, we can generate the complete family of "concentric circles" having the given point as their center.

- Two points on the half-line generate, by the same process, a "ring." By varying the distance between these points, the width of the ring can be as large or as small as we please; the whole plane and the circumference of radius zero are "limiting cases" of such rings.

- The figure formed by two different lines of the star is that of an "angle," the point being its "vertex" and the lines its "side." For every pair of lines there are two angles which "add up" to the whole plane.

- An angle that covers the whole plane once will be formed by just one line. It is convenient to call each side of this line a "lip," so that every line has two lips. In this case, the plane, with the line and its two lips, is <u>the</u> "sheet" of the above-mentioned Riemann surface, which here reduces just to an angle — a complete revolution.

- Two angles may be "added" or "subtracted" by replacing them by another pair having one side in common; this being an addition when the angles are on opposite lips of the common line and a subtraction when they are on the same lip.

- Anything said about one star may be repeated about another star with a different center. The half-lines of two stars can be paired in various ways to generate various loci. For example, when we conceive of each star as generated by rotations in opposite senses, then the intersection of corresponding lines traces the perpendicular bisector of the segment joining the centers of the two stars, with a dramatic, apparent break as the lines pass through positions where they are parallel.

* * *

A star may be used as an introduction to the main notions of trigonometry; this is done in the film Folklore of Mathematics: Trigonometry, further details of which are given in the appendix.

This film starts with a white circle that remains on the screen. A star has its center at the center of the circle. Each half-line cuts the circumference in a point and we can think of this point "describing" the circumference while the half-line rotates to generate the star. This means that we can carry over various properties of the star to the motion of the point moving round the circle.

In particular, the Riemann surface in this case is reduced to the interior of the circle. An angle on one "sheet" of the Riemann surface is increased by a complete revolution as the point moves to the next sheet. A complete revolution is traditionally written 2π. So an angle A on one sheet is more generally defined as A $\pm$ 2π on the n-th sheet. In trigonometry angles are always of this kind: one measures them in one sheet and then the others are determined according to the number of sheets that are traversed.

When it comes to making statements about the film, it is useful to have some agreed frame of reference. For viewers, this can be by gesture, by color, or by agreed, temporary names. In the present context, without the film, we can only rely on some diagrams, which we will refer to in traditional ways. Thus, the horizontal ray to the right and the vertical ray upwards may be chosen as reference lines of a coordinate system, with origin, O, at the center of the circle. The perpendicular from the point P describing the circle to the horizontal diameter meets the latter at S. The point P is associated with the angle made by its corresponding half-line and the horizontal reference line. Calling this angle θ, we can say that on the first sheet, $0 \le \theta \le 2\pi$.

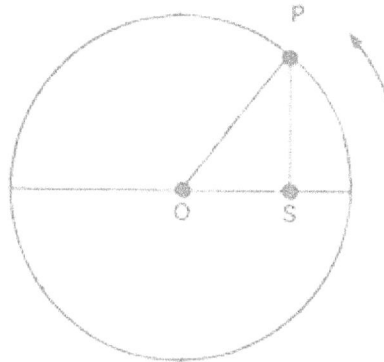

The film shows the point P describing the circumference. The segment OP is a radius and so remains constant in length throughout. This may be taken to be a unit measure. It may be observed that the segments PS and OS vary in length, ranging from zero length to the unit. The changes in their length seem related; as one increases, the other decreases. These changes become clearer when they are related to the changes in the angle θ. The length of the segment PS is written sinθ and read "sine theta;" the length of the segment OS is written cosθ and read "cosine theta." The lengths of these segments are obvious when θ is zero or a complete revolution 2π; and equally obvious at intermediate positions where θ is $\frac{\pi}{2}$, π, or $\frac{3\pi}{2}$.

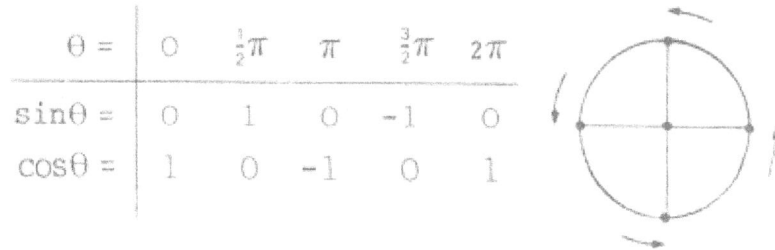

It may be noticed that a particular shape of triangle OPS can appear in four distinct positions, each having the same sine and the same cosine. To distinguish these it is usual to invoke the positive and negative directions of the coordinate system. Thus, for example, in the "third quadrant" the sine PS and the cosine OS are both negative. Certain relationships become immediately obvious from the dynamic image and do not need to be memorized in symbols. For example, the cosine of an angle in the second quadrant is the negative of the cosine of its "supplement":COSθ = COS (π-θ).

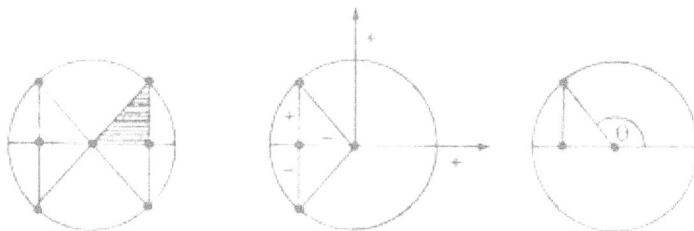

It may be further noticed from the shape of the triangles involved that the cosine of an angle is the sine of the "complement" of the angle. It will also be clear that the sine and cosine functions are "periodic," repeating their values for angles greater than a complete revolution, when the point moves round the circle further times — or, from another point of view, the point moves on to further sheets of the Riemann surface.

* * *

Another important set of relationships is indicated in the film when squares are drawn on the segments OP, PS and OS. The first square remains constant — it is a unit square. The others vary in area from zero to the unit. The changes in their area seem related — as one increases, the other decreases. The area of the square on PS is PS×PS ~ $\sin\theta \times \sin\theta$ which is customarily abbreviated to $\sin^2\theta$ — and similarly for the other square. The images of squares are more suggestive and it seems possible that the two changing squares are complements in the unit square. This is certainly so in the four special positions where one

167

square is zero, and also true for the intermediate positions where both squares are half a unit. These confirm the trigonometric form of Pythagoras' theorem $\sin^2\theta + \cos^2\theta = 1$.

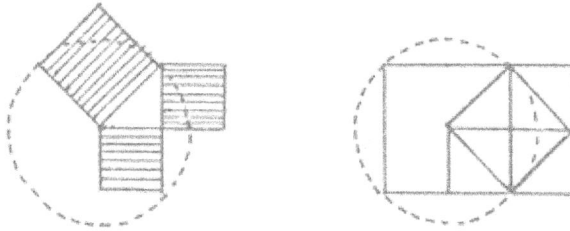

The film also offers an image of another trigonometric function. A line is drawn touching the circle at the endpoint T of the horizontal reference line. The rotating half-line, which cuts the circle at the point P, now also cuts this tangent at a point Q. As P moves round the circle, Q goes upwards along the tangent line, and then dramatically reappears from below, returning to T at the same moment as P does. The segment PT varies in length from zero to infinity and then back to zero. The length of the segment PT is written "$\tan\theta$", read "tangent θ;" here, the name refers to the tangent of the circle on which the function is represented.

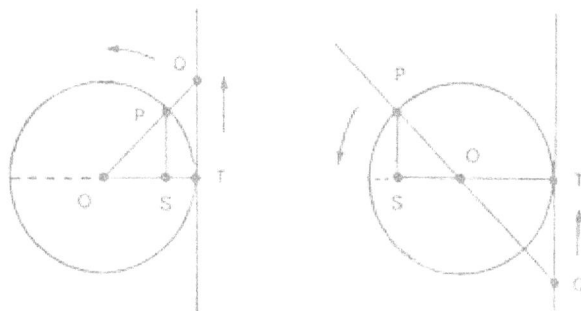

The film then indicates by use of scaling that the tangent function is related to the previous two; in fact, it is their quotient: $\tan\theta = \dfrac{\sin\theta}{\cos\theta}$.

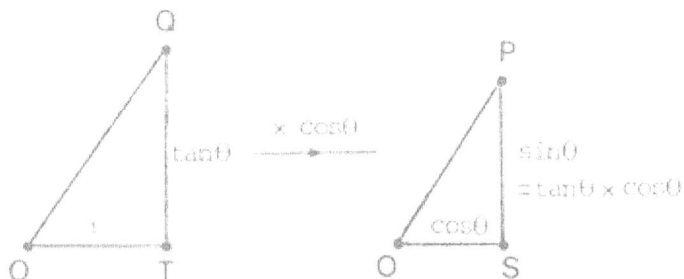

Further sections of the film generate the graphs of the sine and cosine functions. Time is again introduced as a coordinate when the images of sine and cosine segments are allowed to persist as the point moves round the circle. The animation speed with which the point P is taken round the circumference is varied.

This alters the shape of the trigonometric graphs to give waves with differing "wavelengths" and "frequencies."

* * *

It will be evident to secondary school teachers that what can be derived from six minutes of film covers a very great deal of the classical trigonometry required in schools, up to and including the top level classes, just preceding college. It does so in a coherent way which replaces memorization of seemingly interminable symbolic formulae by a single dynamic image which costs very little — a "nothing" — to retain. Moreover, the work presented here requires nothing that is not available to young children, nor does it have to be spread over a number of years.

Trigonometry is one item in the "folklore" of mathematics which can be recast and offered to young children in its entirety.

Summary

Young children are continually investigating their powers of perception and action; they use these powers spontaneously — without having to be taught — in order to elaborate their experience of themselves and of the world. In the first part of this chapter, we have tried to show how this can become part and parcel of mathematical education.

The Geoboards synthesize perception and action. They do so pinpointedly, by requiring separate acts, which yield something that has an existence of its own, in as simple or as complicated a way as the individual child chooses. This can be translated into mathematical language, which from then on triggers images well lodged in the mind.

Another section offered a very different way of enlarging the mathematical experience of students. In the case of film, action is delegated to the animation. This denies viewers the initiative with which they can usually alter the field of perception by using their will. But it provides a new kind of dialogue based on the power of the continuum and the dynamics of the mind.

These are two very different ways of linking awareness of mathematization to the spontaneous ways of knowing belonging to all of us. They both make a mighty difference to how students enter geometry, stay in it during their lessons and afterwards, and integrate it as a "natural" mental behavior with all other behaviors, including verbalization, discipline, contact with challenges, and retention. The resulting mental structures are available for the construction of mathematical mental structures.

Some of the films discussed in this chapter were animated by computer. But computers can be made to be interactive in a different and much more effective way. They have a tremendous future in education, provided that they are no longer used to dispense information. Our description of the approach offered in some films may already suggest that there are ways of involving students in expanding scenes in a number of directions. The great advantage is that they can throw students back on to themselves and make them fully responsible for the steps they take.

We can now say that we <u>know</u> how to educate the mathematician in every one of us, even if only a few of us will become professional mathematicians.

Of course, much more needs to be done. The new science of education needs its scientists. Many workers are required to fill the gaps and to extend the net. We may have to wait a number of years to see, for example, teams of writers of scenarios and programmers coming together to solve the problems of making adequate interactive computer programs. But that so much has been achieved in half a century augurs well for the future.

What is required now is that educators of all kinds make themselves vulnerable to the awareness of awareness, and to mathematization, rather than to the historical content of mathematics. They need to give themselves an opportunity to experience their own creativity, and when they are in contact with it, to turn to their students to give them the opportunity as well.

The three chapters of this volume may contribute to this.

Appendix: Some References

Various books have been written and published by the author about the approach to mathematics that has been described in this volume. These books, and the associated materials are available in the US from Educational Solutions, 95 University Place, New York, NY 10003-4555, and in the UK from Educational Explorers, 11 Crown Street, Reading. These publishers are referred to by the abbreviations Ed Sol and Ed Exp in the following list.

Books

The Science of Education; Part 1, Ed Sol, 1987. This first part of the work contains six chapters, which are devoted to theoretical considerations; it gives references to some further relevant books.

The Common Sense of Teaching Mathematics, Ed Sol, 1974. Numeration and complementarity are discussed in some detail; the rods are presented as a model for the algebra of arithmetic; and there is a final section on teaching. An appendix reprints an article, "A Prelude to The Science of Education," from Mathematics Teaching, 59 (1972).

For the Teaching of Mathematics, 3 vols, Ed Exp, 1963 — a collection of articles published in various journals between 1947 and 1963.

Three books, published at different times, cover between them the period from birth to adolescence and constitute a further background to The Science of Education: The Universe of Babies (Ed Sol, 1973), Of Boys and Girls (Ed Sol, 1975), and The Adolescent and His Self (Ed Exp, 1962); an abridged version of these is now available in one volume, Know Your Children As They Are; A Book for Parents, Ed Sol, 1988.

Computer Software

Visible and Tangible Mathematics: Parts 1 and 2. This is a set of programs on 12 disks for an Apple microcomputer. It constitutes an elementary mathematics course, from numeration to the four operations on the integers, and covers the material discussed in Chapter 10 of this volume.

Educational Solutions Newsletter, vol. XI, no. 3-4 (1982) — an article about the first part of the microcomputer course.

"Operations on Integers," Mathematics Teaching, 114 (1986). This article discusses the second part of the course, which is on two discs and which is devoted to the multiplication and division of integers.

Hands, 16mm film, a sample film made in Toronto in 1971 to illustrate complementarity on the fingers of a pair of hands. Not available.

Cuisenaire Rods

Sets of rods are available in Europe from the Cuisenaire Company, 11 Crown Street, Reading, or in the US — where they are marketed as Algebricks — from Education Solutions.

Mathematics with Numbers in Color, books I-VII, Ed Exp, 1960-65 — a series of textbooks for students that adopts an algebraic approach to elementary arithmetic using the rods (and Geoboards in Book VII).

Gattegno Mathematics, Books 1-5, Ed Sol, 1970 — the American edition of the above series.

<u>Now Johnny Can Do Arithmetic</u>, Ed Exp, 1961 — an introduction to the rods.

<u>For the Teaching of Mathematics</u>, vol. 3, Ed Exp, 1963. This third volume of the collection consists of various articles about the rods and the Geoboards.

<u>Prisms and Cubes</u>, Ed Sol, 1974 — a pamphlet describing the material that is an extension of the rods.

Films

The original Animated Geometry series by J-L Nicolet consisted of 22 short, black and white, silent 16mm films, with titles such as <u>Three Points Determine One Circle</u> and <u>Common Generation of Conics</u>. Four further films in this series were made by C. Gattegno: <u>Extensions of Pythagoras' Theorem</u>, <u>Sections of a Cube</u>, <u>Generation of Some Plane Curves</u>, <u>Sections of a Cone</u>.

<u>For the Teaching of Mathematics</u>, Ed Exp, 1963. The second volume of this collection contains various articles on the original Nicolet films; in particular, various accounts of lessons based on them.

"L'enseignement par le film mathematique," in C. Gattegno et al, <u>Le Material pour L'enseignement des Mathematiques</u>,

Delachaux et Niestlé, 1958. This volume also contains further articles on film by J-L Nicolet and T.J. Fletcher. Was translated into Italian and Spanish but not English.

There are seven films in the new Animated Geometry series; they are listed below. These 16mm films are in color and last from four to six minutes; they were produced by C. Gattegno, named after J-L Nicolet, and computer-animated by J. Chicoine and A. Fourrier. They are available for sale or hire from the addresses given above.

Animated Geometry — new series:
1   Circles in the plane;

2   Angles at the circumference;

3   Common generation of conics;

4   Two circles seen under equal angles;

5   Poles and polars in the circle;

6   Definitions of the right strophoid;

7   Epi- and hypo- cycloids.

<u>Animated Geometry</u>, Ed Sol, 1981 — a pamphlet of notes for teachers on the new series.

The film <u>Folklore of Mathematics</u> was discussed in chapter 12. This is in five parts, each about 4 minutes, available from

Educational Solutions or, in the UK, from the Cuisenaire Company.

"Mathematics and Imagery," <u>Mathematics Teaching</u>, 33 (1965). This article describes briefly the notion of a star and its application in the above file.

<u>Educational Solutions Newsletter</u>, vol. IX, no. 3 (1980) — this issue is mainly devoted to the star and developments from it.

A film, <u>Foundations of Geometry</u>, was produced by C. Gattegno in 1979 for use in a proposed new geometry syllabus. The film lasts 17 minutes and tries to reflect an axiomatic treatment of geometry proposed by the French mathematician, G. Choquet. It is available, with a set of notes, for sale or hire from Educational Solutions.

Geoboards

The various Geoboards described in chapter 12 are available from the previous addresses.

<u>Geoboard Geometry</u>, Ed Sol, 1971. This is a later edition of the pamphlet <u>From Actions to Operations</u>, first published in the UK in 1958.

Students Sheets (duplicating masters) with Notes for the Teacher

These are published by Educational Solutions. Those mentioned here were specially designed by the author and his staff as one way of offering to students many of the awarenesses presented to teachers in Chapters 10 and 11.

<u>For Students of Any Age</u>

Math Mini-Tests* on Computation / Level E^R

    <u>Content:</u> all + and − / × and ÷ tables / 1 digit multipliers / short ÷ no remainder

Math Mini-Tests on Computation / Level I^R

    <u>Content:</u> 2 digit multipliers / ÷ with remainder / long ÷ / fractions, decimals, and percentage.

---

*   The name "Mini-Test" was chosen because these are also part of larger sets of sheets including content that gives students the test-taking skills needed to meet the demands of formal achievement tests, with confidence and ease.

<u>For 6 and 7 Year Olds</u>

Math Mini-Tests / Level P¹ (Basic Starters)

<u>Sheets 2a-2c, 14a-14e</u> on numeration
<u>Sheets 17a-17w</u> on addition & subtraction
(summarized in Eᴿ above)

Math Mini-Tests / Level P² (Basic Starters)

<u>Sheets 9a-9t</u> on addition & subtraction
(summarized in Eᴿ above)
<u>Sheets 26-26L</u> on multiplication tables
(included in Eᴿ above)

www.ingramcontent.com/pod-product-compliance
Lightning Source LLC
Chambersburg PA
CBHW081528220326
41598CB00036B/6365